£7—75
£7/15/-

GAMMA-RAY SPECTROMETRY OF ROCKS

SERIES

Methods in Geochemistry and Geophysics

1. A. S. RITCHIE
CHROMATOGRAPHY IN GEOLOGY

2. R. BOWEN
PALEOTEMPERATURE ANALYSIS

3. D. S. PARASNIS
MINING GEOPHYSICS

4. I. ADLER
X-RAY EMISSION SPECTROGRAPHY IN GEOLOGY

5. THE LORD ENERGLYN AND L. BREALEY
ANALYTICAL GEOCHEMISTRY

6. A. J. EASTON
SILICATE ANALYSIS BY WET CHEMICAL METHODS

7. E. E. ANGINO AND G. K. BILLINGS
ATOMIC ABSORPTION SPECTROMETRY IN GEOLOGY

8. A. VOLBORTH
ELEMENTAL ANALYSIS IN GEOCHEMISTRY
A. MAJOR ELEMENTS

9. P. K. BHATTACHARYA AND H. P. PATRA
DIRECT CURRENT GEOELECTRIC SOUNDING

11. W. ERNST
GEOCHEMICAL FACIES ANALYSIS

Methods in Geochemistry and Geophysics

10

GAMMA-RAY SPECTROMETRY OF ROCKS

BY

JOHN A. S. ADAMS

Professor of Geology, Department of Geology, Rice University, Houston, Texas, U.S.A.

AND

PAOLO GASPARINI

Professor of Earth Physics, Institute of Terrestrial Physics, University of Naples, Naples, Italy

ELSEVIER PUBLISHING COMPANY

AMSTERDAM/LONDON/NEW YORK

1970

ELSEVIER PUBLISHING COMPANY
335 JAN VAN GALENSTRAAT
P. O. BOX 211, AMSTERDAM, THE NETHERLANDS

ELSEVIER PUBLISHING COMPANY LTD.
BARKING, ESSEX, ENGLAND

AMERICAN ELSEVIER PUBLISHING COMPANY, INC.
52 VANDERBILT AVENUE
NEW YORK, NEW YORK 10017

STANDARD BOOK NUMBER 444-40829-0
LIBRARY OF CONGRESS CARD NUMBER 74-103352

WITH 108 ILLUSTRATIONS AND 44 TABLES

PRINTED IN THE NETHERLANDS

To the memory of
Robert A. Welch,
benefactor and citizen
of Houston, Texas

Preface

The gamma-ray spectrometry of rocks is of interest to geochemists, exploration geologists, and radiological health physicists. The geochemists are interested in the various ways the natural radioactive tracers, ^{238}U series, ^{235}U series, ^{232}Th series, and ^{40}K, are distributed amongst the various geochemical domains. Geophysicists are particularly interested in the very substantial contribution that these natural radioactivities make to the heat flowing from the earth's interior. The exploration or economic geologist is especially interested at present in using gamma spectrometry in the search for uranium ore. There is also some secondary interest in using the natural radioactivities as "pathfinders", e.g., uranium in phosphate rock, and uranium together with thorium in placer sands. The radiological health physicist or chemist must view the distribution of fission products against the natural gamma spectrum and this viewing must be made in the geological field context.

The authors have attempted to strike a balance between physical theory and instrumentation on one hand and geologic applications on the other hand. It is hoped that both specialist and student will find this monograph useful and the authors would be grateful to learn of any mistakes or omissions.

The authors wish to acknowledge the support of a NATO Postdoctorate Fellowship (to P.G.) and the continuous support of the Robert A. Welch Foundation, through Grant C-009.

Contents

CHAPTER 1

Physics of Gamma and X-rays

DEFINITION

Gamma and X-rays are highly penetrating electromagnetic radiations characterized by wave lengths in the range of 10^{-7}–10^{-11} cm and, consequently, by frequencies ranging from 10^{17} to 10^{21} sec^{-1}. Fig.1 shows the wave length and frequency ranges of the electromagnetic spectrum; gamma and X-rays are in the short wave length (or high frequency) extreme of the spectrum.

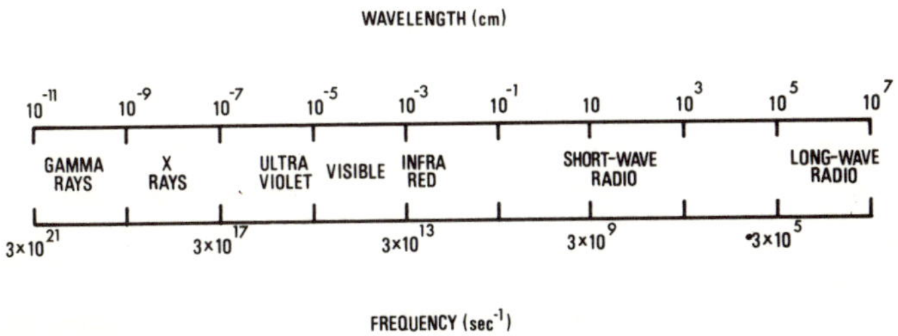

Fig.1. Approximate frequency and wave length of electromagnetic radiations.

Radiant energy is a form of energy propagation in space not involving a measurable mass. Its nature has long been debated and its properties do not readily fit any classical model. Some phenomena, such as diffraction and interference, can be satisfactorily explained by assuming that radiations have a wave nature. The wave theory, however, can only explain in an incomplete way some other characteristics, e.g., the atomic spectra, photoelectric emission, and the intensity of the energy emission from a black body. All these phenomena can be explained satisfactorily by assuming, after Planck, that a body cannot absorb or emit energy in a continuous way, but only as whole multiples of a given amount of energy, called a quantum. The characteristics of the interaction of the electromagnetic radiation with matter, mainly the phenomena of emission of electrons from metals when struck by high frequency radiations (photoelectric emission) and the scattering of such radiations by electrons, required a further modification of the theory. Emission and scattering could be explained only if one assumed that radiation, in addition to being absorbed and emitted in an integral number of quanta, propagates through space as discrete quanta, or photons. Each photon has an energy, E, which is a function of the characteristic frequency (or wave length) of the radiation, i.e.:

$$E = \mathrm{h}\nu = \mathrm{hc}/\lambda \tag{1.1}$$

where ν is the frequency and λ the wave length of the radiation; c is the velocity of light, and h is the Planck constant. When the frequency is expressed in sec^{-1} and the energy expressed in ergs, the Planck constant has the value:

$$\mathrm{h} = 6.62554(\pm 0.00015) \cdot 10^{-27} \text{ erg sec}$$

Gamma and X-ray energies are usually expressed in electron volts (eV). One electron volt is the energy acquired by a charged particle carrying unit electronic charge when it is accelerated through a potential difference of one volt. One eV is equivalent to $1.602 \cdot 10^{-12}$ erg. One eV is associated through the Planck constant with a photon of wave length 1.2395 μ. The energies of natural X and gamma rays are of the order of 10^3–10^6 eV, and they are usually expressed in keV (10^3 eV) or in MeV (10^6 eV).

X and gamma rays are classically differentiated according to their wave lengths: electromagnetic radiations with frequencies between 10^{19} and 10^{21} sec^{-1} (corresponding to wave lengths between 10^{-9} and 10^{-11} cm and to energies between 40 keV and 4 MeV) are called gamma rays; those with frequencies ranging from 10^{17} to 10^{19} sec^{-1} (corresponding to wave lengths from 10^{-7} to 10^{-9} cm and to energies from 40 eV to 40 keV) are called X-rays. It has been found more meaningful and unambiguous to distinguish X from gamma rays on the basis of where they originate. From this standpoint, gamma rays are the radiations originating within the

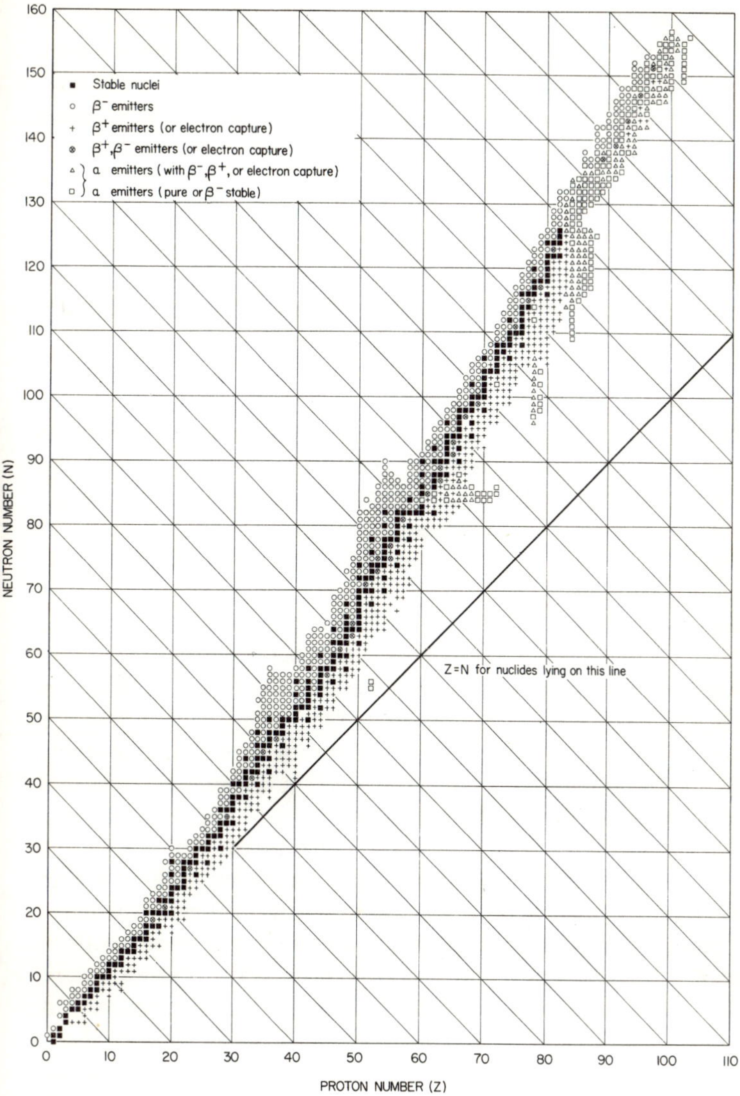

Fig.2. Chart of isotopes. (Drafted from the data of LEDERER et al., 1967.)

atomic nucleus, while X-rays originate in the electron cloud surrounding the nucleus. When gamma and X-rays are defined in this way, their frequency (or energy) ranges overlap considerably. Gamma rays having energies as low as a few keV have been in fact reported (e.g., ^{169}Er, 8 keV, and ^{81}Kr, 12 keV), whereas K X-rays from transuranium elements have energies as high as 140 keV.

GAMMA EMISSION

Gamma-ray emission is one way a nucleus gives off the energy excess it can have as a result of radioactive decay or induced nuclear reactions. Radioactive decay consists of the emission or capture of elementary or composite particles with consequent transformations into daughter nuclei characterized by different atomic numbers and in some cases by different mass numbers.

Fig.2 is a plot of number of neutrons vs. number of protons for all known nuclei. It has been drawn from the data of LEDERER et al. (1967). The neutron/proton ratio for light stable nuclei (with atomic number up to 20) is nearly one; from there it increases up to a value slightly higher than 1.5 for stable nuclei with mass number of about 80. Other general systematics include: (*1*) nuclei with an even number of protons and an even number of neutrons are the most abundant and are generally more stable; (*2*) nuclei with odd protons and odd neutrons (i.e., odd Z and even A) are the least stable; (*3*) nuclei in which either protons or neutrons are even (odd–even nuclei) are intermediate in stability.

When a nucleus has a neutron/proton ratio too high for stability it tends to decrease the number of excess neutrons and to increase the number of protons by transformation of a neutron into a proton, which proton is retained within the nucleus, and with the emission of an electron (β^-)–anti-neutrino pair. When a nucleus has a neutron/proton ratio too low for stability, it can decrease the number of excess protons by one of three processes: alpha emission, positron (β^+) emission, and orbital electron capture (K-capture). The alpha emission consists of the expulsion of an He^{2+} nucleus from the parent nucleus, and, because of the high energy involved, is relatively infrequent. It occurs mainly in nuclei with mass number higher than 210; only a few alpha emitters with lower mass number are known. Positron emission consists of the transformation of a proton within the nucleus to a neutron, positron, and neutrino, with the emission of the positron–neutrino pair; consequently the nucleus has one more neutron and loses one positive charge. Orbital electron capture consists of the capture by the nucleus of an extranuclear electron with the emission of a neutrino; in most cases this electron belongs to the K or innermost shell of the atom (the name K-capture is therefore often used). Positron emission and orbital electronic capture are alternative modes of decay for many nuclides and have the same effect, i.e., the atomic number decreases one unit without any major change of mass in the nucleus. The energies of some of the above described processes can be shared in various continuous ways among three particles. Each kind of these particles can be emitted with non-discretely changing energies; as a consequence their energy spectrum is a continuum.

Energy requirements set a lower limit for the occurrence of positron emission, i.e., the atomic mass difference between parent and daughter must be higher than twice the rest energy of the electron (0.511 MeV). Starting from this value, the relative probability of positron emission increases with increasingly available decay energy. As a broad generalization, it can be therefore stated that the electron capture/positron emission ratio for a given nuclide, when enough energy for positron emission is available, decreases with increasing energy. Conversely the probability of orbital electron capture increases for increasing atomic number. Additional types of decay, e.g., neutron and proton emission, have not been observed frequently in rocks.

In many cases, nuclides can decay by two of these processes, for instance K[1] capture and β^- emission (e.g., ^{40}K[1]) or α and β^- emission (e.g., ^{212}Bi and ^{214}Bi). These nuclides are said to undergo "branched decay". The relative probability of each type of decay is usually called "the branching ratio".

---

[1] The symbol "K" is used in nuclear physics for the innermost electron shell (see pp.10–12); the symbol K (for Kalium) is also used in chemistry as the symbol for potassium.

Emission of gamma rays often accompanies each of the above described processes. Nuclei decaying by any process have been observed to emit primary particles which frequently are not all of the same energy, but can be gathered into two or more discrete groups. Gamma rays are commonly associated with this kind of nuclei. A typical example is ^{212}Bi. This nuclide is a member of the ^{232}Th natural series and undergoes branched decay, i.e., it decays by α emission to ^{208}Tl and by β^- emission to ^{212}Po. Let us consider, as an example, only the alpha decay. The emitted alpha particles can be gathered into different energy groups. Table I

TABLE I

DIFFERENT ENERGETIC GROUPS OF ALPHA PARTICLES FROM ^{212}Bi

Alpha particles energy (MeV)	*Number of α particles per hundred α disintegrations*	*Alpha disintegration energy (MeV)*	*Difference of alpha disintegration energy from that of the most energetic group*
6.084	27.2	6.201	0
6.044	69.9	6.161	0.040
5.762	1.7	5.873	0.328
5.620	0.15	5.728	0.473
5.601	1.1	5.708	0.493
5.481	0.016	5.585	0.616

summarizes the number of particles of each group emitted for a hundred alpha disintegrations, the energies of the alpha particles, and the alpha disintegration energies. The latter are different from the energies of alpha particles in this respect: when an alpha particle is emitted, the principle of conservation of momentum requires that the residual nucleus recoils, so that:

$$Mv = M_n v_n \tag{1.2}$$

where M and v are the mass and velocity of the alpha particle and M_n and v_n are the mass and energy of the recoiling nucleus. Where alpha particles are considered, the relativistic effect can be neglected so that the whole energy of the disintegration process (E_α) is given by:

$$E_\alpha = \tfrac{1}{2}Mv^2 + \tfrac{1}{2}M_n v_n^2 \tag{1.3}$$

and, upon substitution with eq.(1.2):

$$E_\alpha = \tfrac{1}{2}Mv^2(1 + M/M_n) \qquad (1.4)$$

Therefore the alpha disintegration energy differs from the alpha particles energy by a factor of $(1 + M/M_n)$. The differences among the alpha disintegration energy relative to the most energetic alpha emission (i.e., 6.084 MeV) and those relative to the other emissions are given in Table I. Some of the gamma emissions from ^{212}Bi can be gathered into the following energy groups: 0.040, 0.144, 0.164, 0.288, 0.328, 0.432, 0.452, and 0.472 MeV. A comparison with the disintegration energies reported in Table I shows that they match closely the energetic differences corresponding to different groups of alpha emissions. An hypothesis explaining the meaning of the energetic spectra displayed by primary particles and the way such spectra are bound to gamma emission has been put forward by Gamow in 1930.

Today the existence of different energy levels in the nucleus is a major hypothesis of nuclear physics. It is assumed that a nucleus is ordinarily in a state corresponding to the lowest energy level (ground state), but that it can be found for short times in states characterized by higher energies (excited states). When the above reported spectrum is interpreted from this point of view, the most energetic disintegration can be expected to correspond to a transition from the ground state of the parent nucleus (^{212}Bi) to the ground state of the daughter nucleus (^{208}Tl). All the other alpha transitions are assumed to start from the ground state of the parent nucleus, but they leave the daughter nucleus in an excited state because not all the energy available from the decay has been spent in the alpha transitions. The excited nucleus emits the energy excess by emitting a gamma ray or more gamma rays in cascade that bring the nucleus to lower and lower excited levels. Each gamma ray can be matched to the difference of energies between two nuclear energetic levels, and the energies of primary particles and gamma rays can be worked out in such a way as to build up an energy decay scheme.

The usual way of representing decay schemes is shown in Fig.3. Decay modes leading to lower atomic number (e.g., alpha and positron emission, electron capture) are represented as in Fig.3;

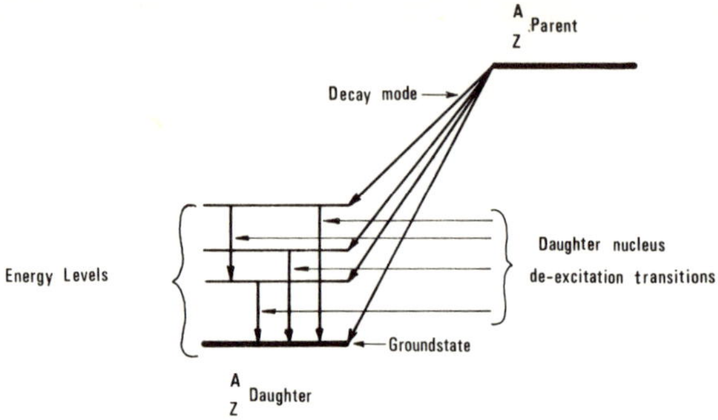

Fig.3. Graphical representation of a decay scheme.

for β^- decay the daughter nucleus is placed on the right-hand side as in Fig.4. The energy levels are represented by horizontal lines, whereas tilted arrows from parent to daughter energy levels show

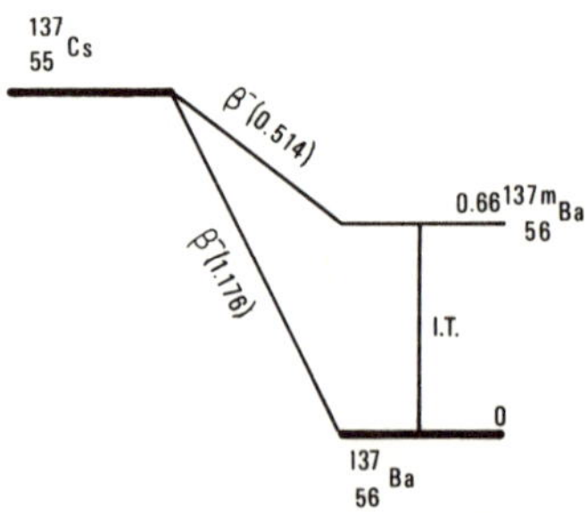

Fig.4. Decay scheme of ^{137}Cs.

the energetically different decay modes. Vertical arrows among energetic levels represent energy (electromagnetic) transitions within the same nucleus, not implying any isotopic transformation. Additional data such as half lives, energies of primary and gamma radiations, and other nuclear characteristics are usually reported in decay schemes. The number of primary particles or γ-rays of a given energy emitted per hundred disintegrations of the parent nucleus (absolute intensities) are also reported.

Given that gamma rays arise from the excess nuclear energy

from a primary decay, it follows that they are emitted from the excited daughter and not from the parent, as is loosely stated. For instance, in the above described case, gamma rays usually ascribed to ^{212}Bi must be in effect emitted from excited ^{208}Tl. The study of the internal conversion (IC) phenomena has shown that this indeed is the case. It has, in fact, been observed that gamma emitting nuclides often display a β^- spectrum that, being a line spectrum, is easily distinguished from the continuous spectrum of electrons emitted in β^- decay. The characteristic energies of the line spectrum have been found to correspond closely to those of the gamma rays emitted from the nucleus, when the binding energies of orbital electrons are accounted for. It was first thought that the electron line spectrum was due to interaction of nuclear gamma rays with atomic electrons. Gamma rays from the nucleus were assumed to strike such electrons giving them all of their energy. The electrons were consequently ejected from their shell with an energy equal to that of the impinging gamma rays minus their binding energy. This interpretation has been shown to be incorrect. The internal conversion has to be considered as an alternative process of de-excitation of the nucleus in that the expulsion of atomic electrons is due to a direct interaction with the nuclear field. After the orbital electron has been ejected, the electron shells are reconstituted, the vacancy is filled, and X-rays are emitted. The emitted X-rays have always been found to be of the energy characteristic of the daughter nucleus. Energy involved in internal conversion is the same as that involved in gamma emission; both phenomena represent the same energetic transitions of the nucleus and, therefore, gamma rays must be emitted from the daughter nuclei excited after the primary decay. Other experiments have also confirmed the energetic equivalence of gamma rays and internal conversion electrons.

The relative probability of occurrence of gamma-ray emission or internal conversion de-excitation for a given energy level of a given nuclide is expressed as the "internal conversion coefficient". It is the ratio of the emission rate of conversion electrons to that of gamma rays. The internal conversion coefficient increases rapidly with decreasing de-excitation transition energy and with increasing number of electron shells. Therefore internal conversion is an important process for heavy radionuclides where low energy transitions occur. Tables of internal conversion coefficients are reported by SLIV and BOND (1965).

Gamma emission follows, in most cases, very shortly after the primary radioactive decay. Many examples, however, have been found of delayed gamma emission and delays of hours or less have been observed for some nuclides. Nuclei displaying such characteristics are usually called "nuclear isomers" and they are indicated by an "m" after the mass number (e.g. ^{137m}Ba). This phenomenon is called "nuclear isomerism". Nuclear isomers are metastable excited states of a given nucleus and therefore all their chemical and other properties, but the radioactive, are the same as the corresponding unexcited nuclei. It has been observed that isomeric transitions are often the cause of the breaking of a chemical bond. A classical case of nuclear isomerism and the only one found in natural radioisotopes is that of ^{234m}Pa. As shown in Fig.59, both ^{234m}Pa and ^{234}Pa derive by β^- decay of ^{234}Th. ^{234m}Pa (half life = 1.18 min) decays by β^- emission to ^{234}U or by gamma emission (isomeric transition or I.T.) to ^{234}Pa which in turn disintegrates to ^{234}U by a less energetic β^- emission with a half life of 6.7 h. Many cases of nuclear isomerism are found among artificial isotopes, a typical one being that of ^{137m}Ba illustrated in Fig.4. ^{137m}Ba is produced by β^- decay from ^{137}Cs, which can also decay by a more energetic β^- transition to ^{137}Ba. ^{137m}Ba has a half life of 2.3 min. An hypothesis for the origin of the nuclear isomerism has been proposed by Weizsäcker. He suggests that delayed gamma emission can be a consequence of very different angular momenta of a nucleus in its highly excited and low energetic states.

X-RAY EMISSION

The extranuclear part of the atom consists of electrons ordered in different energetic levels, usually referred to as orbits or shells. As a rough classification, orbits or shells are distinguished as K, L, M, etc., in order of decreasing binding energy, i.e., from the innermost to the outer shells. A more detailed structural study shows that, although the K shell cannot be subdivided furtherly, the L shell can be divided into three subshells (L_I, L_{II}, L_{III}), the M into five subshells etc. When a vacancy is formed in an inner shell, it is filled within a femtosecond (10^{-15} sec) by an electron coming from the outer shells.

The most usual energy transitions occurring when a vacancy is formed in the innermost (K) shell are indicated in this way:

K_{α_1}: K—L_I K_{α_2}: K—L_{II}

K_{β_1}: K—M_{III} K_{β_2}: K—N_{III}

K_{β_3}: K—M_{II} K_{β_4}: K—N_{II}

K_{β_5}: K—M_{IV}

An energy equal to the energy difference of the two shells involved in the transition is consequently liberated and may be emitted as X-rays. The liberated energy can also be transferred to an orbital electron, which is then expelled from the atom. Electrons of this origin are called "Auger electrons" and display a line spectrum that can be easily distinguished from the continuous spectrum of nuclear electrons. Their energy is equal to the energy difference between the two shells minus the electron binding energy. The ratio of the number of emitted X-rays to that of primary vacancies formed in a given shell of a given atom is called "fluorescence yield" of that shell. The ratio of the number of Auger electrons to that of primary vacancies in a given shell of a given atom is called "auger yield" of that shell of the atom. Fig.5 reports the variation of the K fluorescence yield with the atomic number.

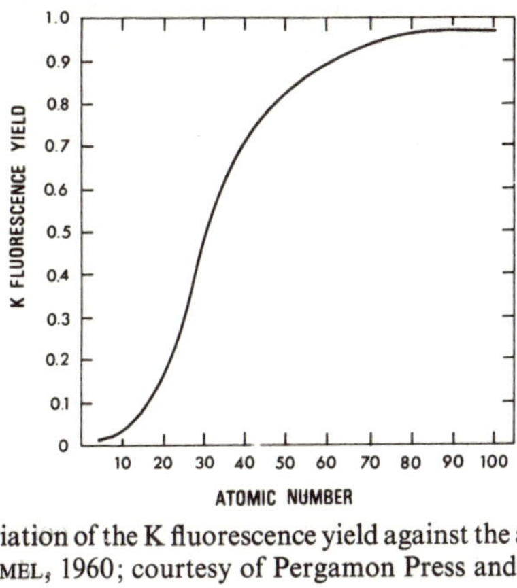

Fig.5. Variation of the K fluorescence yield against the atomic number. (CROUTHAMEL, 1960; courtesy of Pergamon Press and of the author.)

Internal conversion phenomena and orbital electron captures are causes of primary vacancies in the inner shells. The probability that in internal conversion processes the conversion electron belongs to the L or to the K shell (L/K ratio) depends, in a complicated way, on the atomic number of the element, the excitation energy of the nucleus, and other factors. When a nuclide decays by orbital electron capture, K electrons are in most cases captured. The L/K ratio is very low for low atomic number but increases with increasing atomic number up to about 0.15 for nuclei with $Z = 80$.

Another cause of primary vacancies can be the interaction of inner orbital electrons with gamma rays from outside the atom. The gamma ray gives practically all of its energy to an inner electron which can be consequently expelled from the atom, if it receives enough energy to break its atomic bond.

All the X-rays emitted in the above described processes display a line spectrum characteristic of the element and of the orbital shells involved. X-rays can also display a continuous spectrum. When electrons approach the electrostatic field of a nucleus, the electrostatic interaction produces an acceleration and a deflection of the electrons from their original paths, with a consequent loss of energy by radiation. The emitted radiation is usually in the frequency range of low energy gamma rays and X-rays and is continuous because of the statistical nature of the electron energy loss, with an upper limit determined by the available energy. The emitted radiation is usually called bremsstrahlung (braking radiation). The fractional energy loss by bremsstrahlung of an electron interacting with a nuclear field increases with increasing energy of the electron and with increasing proton number of the nucleus.

ABSORPTION OF GAMMA AND X-RAYS

The absorption of photons by matter has characteristics very different to that of the absorption of charged particles. When a beam of monoenergetic alpha particles passes through an absorbing medium, each single particle is gradually slowed down by electrostatic interactions or by collision with ions, electrons, nuclei and atoms. Finally, the particle captures two electrons and comes to rest. The kinetic energy of the whole beam consequently decreases continuously along the path, but the number of particles forming the beam remains practically constant until, after the same average path in the absorber, all of them are stopped. The length of the path is a function of the nature and initial energy of the particle and of the nature of the absorber; it is usually referred to as "range".

The β^- particles are also absorbed in a rather similar way and have a more or less definite "range" when they pass through a given absorber. The above described absorption mechanism is, however, not strictly applicable to β^- rays, mainly because of their smaller mass.

When a beam of monoenergetic parallel photons passes through an absorber, each single photon of the beam either disappears after giving off all its energy in a single event or is scattered away from the beam. Therefore the number of photons in the beam decreases continuously as the beam penetrates deeper and deeper into the absorbing medium. If I is the intensity of the beam, defined as:

$$I = \varphi \mathrm{h}\nu \tag{1.5}$$

where φ is the number of photons crossing the unit area in the time unit (flux of photons) and $\mathrm{h}\nu$ the energy of each photon when the beam is passed through a thickness $\mathrm{d}x$ of absorber, the intensity decreases by an amount $\mathrm{d}I$:

$$\mathrm{d}I = -\mu I \,\mathrm{d}x \tag{1.6}$$

where μ is a constant for a given absorbing medium and a given photon energy, called the "absorption coefficient". When eq.(1.6) is integrated, we get:

$$I = I_0 \exp(-\mu x) \tag{1.7}$$

where I_0 is the intensity of the beam when $x = 0$.

When x is expressed in cm and the intensity analogously in cm–g–sec units, μ is consequently expressed in cm^{-1}. However, it has been found convenient to sometimes express the thickness by multiplying the linear dimension of the absorber by the density; it is therefore expressed in $\mathrm{g/cm}^2$. The absorption coefficient is consequently given in cm^2/g. The convenience of using this unit arises from the difficulty of getting accurate determinations of the thickness for very thin absorbers. In this case it is much easier and more accurate to determine the weight and the cross section of the absorber. This absorption coefficient is called mass absorption coefficient and is indicated by μ/ρ. Additional ways of giving the thickness of the absorbers are electrons/cm^2 (electron cross section) and atoms/cm^2 (atomic cross section). The absorption coefficients are then given in cm^2/electron (μ_e) and cm^2/atom (μ_a). The relationships among such units are:

$$\begin{aligned} \mu_a &= Z\mu_e \\ \mu/\rho &= N(Z/A)\mu_e = (N/A)\mu_a \qquad (1.8) \\ \mu &= \rho N(Z/A)\mu_e = (\rho N/A)\mu_a \end{aligned}$$

where Z and A are the atomic and mass numbers of the absorber and N the Avogadro number.

The absorbing properties of a given material can also be expressed in terms of half-thicknesses. The half-thickness can be defined as the thickness of a given absorber needed for reducing the intensity of a beam of parallel monoenergetic photons to half its initial value. From eq.(1.7) we get:

$$\log \tfrac{1}{2} = -\mu x_{\frac{1}{2}} \log e \qquad (1.9)$$

and:

$$\mu = 0.693/x_{\frac{1}{2}} \qquad (1.10)$$

Fig.6 reports the half-thicknesses of a few elements of widely different atomic number.

The absorption of photons can occur by three main processes: (*1*) the photon gives all its energy to a bound electron and disappears (photoelectric effect); (*2*) the photon gives a part of its energy to an electron and is scattered away from the beam (Compton scattering); (*3*) the photon interacts with the electrostatic field of a nucleus giving rise to the formation of a positron–electron pair and disappears (pair production). Some other processes may also occur, such as coherent scattering by atoms or incoherent scattering by free electrons (Thomson scattering), but they are of relative minor importance.

The probability of a given absorption process is often expressed in terms of cross section. The cross section can be considered as the target area presented by a particle to the incident beam of photons.

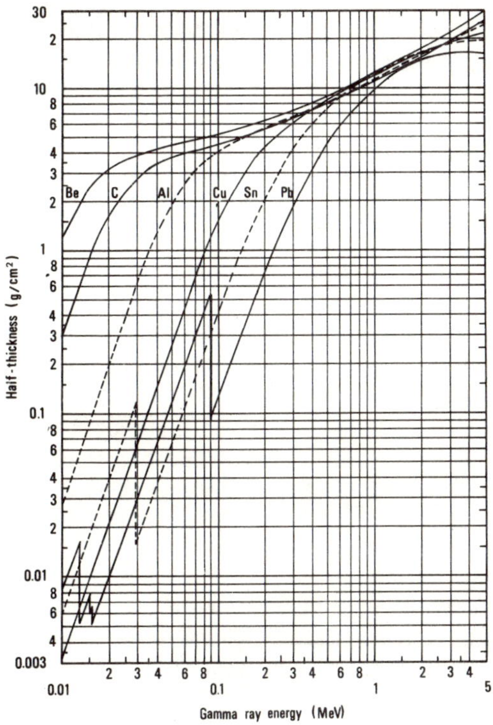

Fig.6. Half-thicknesses of Be ($Z = 4$), C ($Z = 6$), Al ($Z = 13$), Cu ($Z = 29$), Sn ($Z = 50$), Pb ($Z = 82$) as a function of gamma energies. (After WAPSTRA et al., 1959; courtesy of North-Holland Publ. Co.)

It is usually given in units of 10^{-24} cm^2 (barns). Concluding, the total absorption coefficient (μ) for a beam of monoenergetic photons crossing a given absorber is practically the sum of the photoelectric absorption coefficient (μ_{ph}), of the Compton absorption coefficient (μ_c) and of the pair production absorption coefficient (μ_{pp}):

$$\mu = \mu_{ph} + \mu_c + \mu_{pp} \tag{1.11}$$

Absorption coefficients, expressed in terms of cm^2/atom, for photoelectric, Compton, pair production and total absorption for different absorbers are reported in Appendix 1. The relative importance of the three major types of X and gamma-ray interaction with matter for different photon energies and for absorbing media of different atomic numbers is schematically shown in Fig.7.

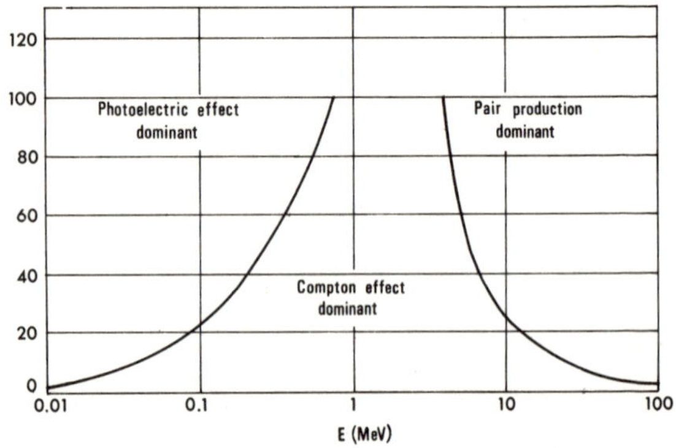

Fig. 7. Regions of domain of photoelectric, Compton and pair production absorption processes. (After EVANS, 1955; courtesy of McGraw-Hill Book Co.)

Photoelectric absorption

In this process the photon transfers all its energy to a bound electron (that the electron must be bound is due to requirements of conservation of momentum), which is expelled from the atom with a kinetic energy, E:

$$E = h\nu - E_b \tag{1.12}$$

where E_b is the binding energy of the electron and $h\nu$ the energy of the incident photon. A threshold value exists for the ejection of electrons; it can occur only if the energy of the incident photon is higher than E_b.

Photoelectric absorption takes place mainly in the inner shells of the atom; about 80% of it occurs in the K shell and a large portion of the remaining 20% in the L shell. After the electron has been ejected, the atom reorganizes and consequent emission of X-rays and Auger electrons occurs.

The cross section for photoelectric absorption depends, in a rather complicated way, on the atomic number of the absorber and on the energy of the photon. When the energy of the photon is so low that special relativistic considerations need not be applied, but high enough to allow the binding energies of the electrons to be neglected, the cross section for photoelectric absorption occurring in the K shell increases approximately with Z^5 and decreases with $(h\nu)^{7/2}$.

Photoelectric absorption is therefore a predominant process at very low energies, but its relative probability decreases rapidly for increasing energy of the photon. For the same photon energy, the probability of photoelectric absorption is higher the higher the atomic number of the absorber (see Fig.7).

Compton scattering

When a gamma ray interacts with a free or loosely bound electron, it can transfer part of its energy and is also scattered away at an angle θ. This process occurs mainly with outer electrons of the atoms and, therefore, is not accompanied by significant emission of X-rays or Auger electrons. As a consequence of the partial energy loss, the scattered photon has a lower frequency (or a higher wave length) as shown in Fig.8. The energy $(h\nu)_\theta$ of a photon scattered at an angle θ can be approximately evaluated with the equation:

$$(h\nu)_\theta = \frac{(h\nu)_0}{1 + \frac{(h\nu)_0}{mc^2}(1 - \cos\theta)} \qquad (1.13)$$

where $(h\nu)_0$ is the initial energy of the photon and mc^2 the rest energy of the electron (0.511 MeV). Eq.(1.13) shows that $(h\nu)_\theta$

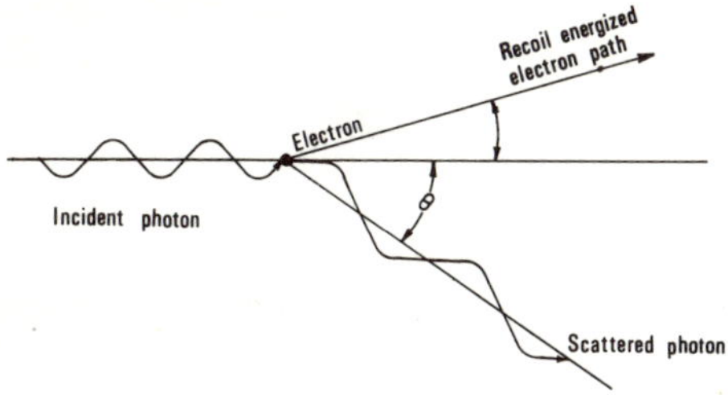

Fig.8. Schematic representation of the Compton scattering.

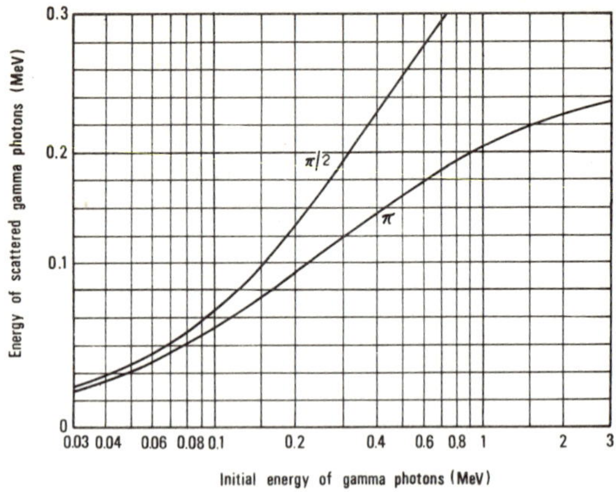

Fig.9. Variation of the energies of 90° (π/2) and 180° (π) scattered photons as a function of the initial energy. (After Wapstra et al., 1959; courtesy of North-Holland Publ. Co.)

decreases with increasing θ down to a minimum, where $\theta = 180°$, i.e., when the photon is backscattered. Furthermore, it shows that the energy of the backscattered photons is only a function of the initial energy of the photon. Fig.9 illustrates the variation of the energy of 180° and 90° scattered photons as a function of their initial energy. For very high initial energies the backscattered

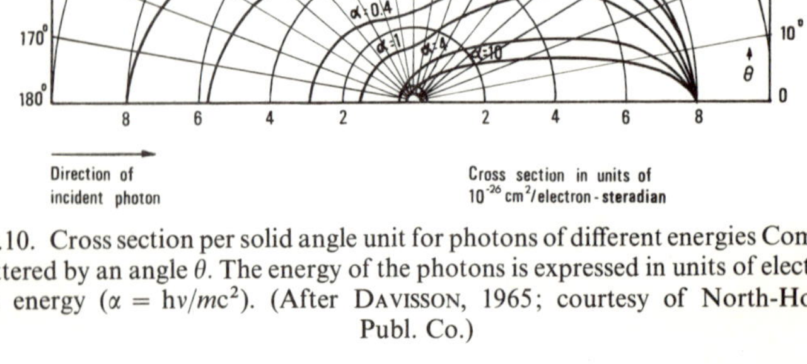

Fig.10. Cross section per solid angle unit for photons of different energies Compton scattered by an angle θ. The energy of the photons is expressed in units of electron's rest energy ($\alpha = h\nu/mc^2$). (After Davisson, 1965; courtesy of North-Holland Publ. Co.)

photon energy approaches a maximum value of about 0.255 MeV.

Where the absorber is very thin, the scattered photon has a high probability of escaping out of it, but such a probability decreases with increasing thickness of the absorber. In thick absorbers, e.g., several centimeters of rock, the scattered photon has, of course, an increasing probability to transfer a great part or all of its energy by repeated Compton scattering and eventual photoelectric absorption.

Fig.10 shows the cross section for scattering into a unit solid angle by an average scattering angle θ for photons of different initial energy. The probability of forward scattering increases with increasing initial energy and forward scattering is practically dominant for initial energies above 2.0 MeV.

The energy absorbed by the recoiling electron (E_r) increases consequently with increasing scattering angle up to a maximum value at $\theta = 180°$ where it is given by:

$$(E_r)_{180} = (h\nu)_0 - (h\nu)_{180} = \frac{(h\nu)_0}{1 + mc^2/2(h\nu)_0} \qquad (1.14)$$

Klein and Nishina derived an equation allowing the evaluation of the probability that a photon having initial energy $(h\nu)_0/mc^2(\alpha)$ is Compton scattered when a collimated beam passes through 1 cm of absorber. The equation is based upon the assumption that colliding electrons are free and at rest. Failure of this assumption gives rise to slightly higher than observed cross sections. The Klein–Nishina formula is very useful for most practical purposes:

$$\sigma_c = \tfrac{3}{4}\varphi_0 \left\{ \frac{1 + \alpha_0}{\alpha_0^2} \left[\frac{2(1 + \alpha_0)}{1 + 2\alpha_0} - \frac{\ln(1 + 2\alpha_0)}{\alpha_0} \right] + \right.$$

$$\left. + \frac{\ln(1 + 2\alpha_0)}{2\alpha_0} - \frac{1 + 3\alpha_0}{(1 + 2\alpha_0)^2} \right\} \text{cm}^2/\text{electron} \qquad (1.15)$$

The constant φ_0 is the cross section for scattering of low energy photons by free electrons at rest (Thomson scattering), and is equal to $6.652 \cdot 10^{-25}$ cm^2. Compton scattering cross sections have been evaluated with the Klein–Nishina equation by NELMS (1953).

Compton absorption probability decreases with increasing initial energy of the photon but at a much slower rate than photoelectric absorption.

Pair production

The solution of the Dirac electron equation, facing the problem of the relativistic treatment of the electron, required the existence of electrons in two different sets of energetic states[1]: negative and positive. Positive energy states are those considered in classical physics, and include the rest energy; negative states have no correspondence in classical physics and can hardly be visualized. Positive energy states must be higher than mc^2 (the rest mass of the electron) and negative energy states must be lower than $-mc^2$. No electrons can therefore exist in energy states between $-mc^2$ and mc^2, i.e., an energy gap of $2mc^2$ exists. In the theory of quantum mechanics, electrons in negative energy states can acquire enough energy to jump over the $2mc^2$ energy gap and thus transfer themselves to a positive energy state. The inverse process could not be excluded on a theoretical basis but, as it would be contrary to experimental evidence, Dirac assumed that it cannot occur because negative energy states are completely occupied by electrons so that, from the Pauli exclusion principle, no transition to such states can be possible.

When an electron acquires enough energy to pass from a negative to a positive energy state, it leaves a hole in the continuum distribution of negative charges present in the negative energy states; this hole will appear as a positive charge of the same magnitude as the negative charge transferred away. The transferred electron will appear as one more negative charge in the positive energy states.

Some years after the publication of the Dirac theory, a charged particle having the same characteristics of the electron but opposite charge, was identified in the cosmic radiation and given the name "positron". The positron can be identified as the hole in the continuum of negative energy states. The process predicted by Dirac leads therefore to the formation of a pair of particles, electron

[1] It should be noted that they are energy states, and have nothing to do with the electric charge of electrons.

and positron, having equal mass and opposite charge. This process is usually referred to as pair production.

A gamma ray incident on an electron in its negative state of energy can give it all its energy and, when it is higher than $2mc^2$ (1.02 MeV), the electron acquires enough energy to jump over the energy gap; the gamma ray disappears and an electron–positron pair is so formed. Requirements of conservation of the energy and momentum of the system allow this process to occur only in the electric field of a nucleus.

After the pair has been formed, the positron is slowed down by interaction with electrons and its energy gradually decreases down to the rest energy. Then it annihilates by combination with an electron. The rest energy of the two mutually annihilating particles is transformed into radiant energy; two gamma rays of 0.511 MeV each are consequently emitted in nearly opposite directions. The cross section for pair production by gamma rays increases with increasing energy, starting from 1.02 MeV, and it is directly proportional to Z^2. Pair production is, however, a far less important absorption process than photoelectric and Compton in the energy range of natural gamma rays. It becomes comparatively important for heavy absorbers, e.g., lead.

PHOTODISINTEGRATION

When a photon has an energy higher than the binding energy of some nucleons in a nucleus it interacts with, all its energy can be transferred to a nucleon and the latter can be expelled. A nuclear transmutation is in this way induced by photons. It is called "photodisintegration" and indicated as (γ, n), (γ, p), (γ, α) according to the nature of the expelled nucleon. Threshold energies for (γ, n) reactions are generally lower than (γ, p) reactions because of the Coulomb barrier working against the ejection of charged particles from the nucleus. Thresholds for (γ, α) reactions are, of course, still higher. Threshold values for (γ, n) reactions are usually higher than 5 MeV; therefore they cannot be generally induced by natural gamma rays from rock whose maximum energy is practically 2.615 MeV (^{208}Tl).

Two isotopes occur in nature which have a loosely bound neutron: ^{2}H and ^{9}Be. Threshold energies for (γ, n) reactions are 2.23 MeV (^{2}H) and 1.67 MeV (^{9}Be). The reactions are:

$$^{2}_{1}H + \gamma = ^{1}_{1}H + ^{1}_{0}n$$

$$^{9}_{4}Be + \gamma = ^{8}_{4}Be + ^{1}_{0}n$$

Both reactions have a geological application. The former can be used for the identification of acquifers in wells or bore-holes, by bombarding the walls with a suitable gamma source and recording the neutron emission rates. The latter can be used in the same way for identifying Be minerals and ores.

For a general treatment of radioactivity the reader is referred to the textbooks by GLASSTONE (1958), KAPLAN (1963), SEGRÈ (1965). The interaction of gamma rays with matter has been discussed more extensively by DAVISSON (1965).

CHAPTER 2

Detectors

INTRODUCTION

The quantitative detection of X and gamma photons is based upon their interactions with matter. Any instrumentation for their detection consists essentially of two parts: a detector and an amplifying–counting system. The detector is the part where the interaction of photons with the matter occurs and an output pulse is generated, whereas the amplifying and counting system increases the detector's output pulse, counts the events and, eventually, discriminates them according to their energies (see Chapter 3 for amplifying–counting).

Before describing the main types of detectors, some terms that will be used extensively in describing the performances of the detectors should be defined.

Intrinsic efficiency of a detector is the ratio of the detected photons to those reaching the sensitive volume of the detector.

Total efficiency is the ratio of the detected photons to those emitted from the source under consideration.

The *geometry* is the fraction of total solid angle, 4π, that the detector subtends relative to a point source.

Where the pulse height distribution resulting from the absorption of a monoenergetic beam of photons is considered, i.e., where an X or gamma-ray spectrometer is used, a typical spectrum is obtained, such as that shown in Fig.11, which is characteristic of 0.4–1.3 MeV gamma rays. The spectrum essentially consists of a Gaussian peak (full energy peak) corresponding to the absorption of all the energy of photons within the detector and of a lower energy continuum mostly resulting from single or multiple Compton interaction. The ability of spectrometric instrumentation to

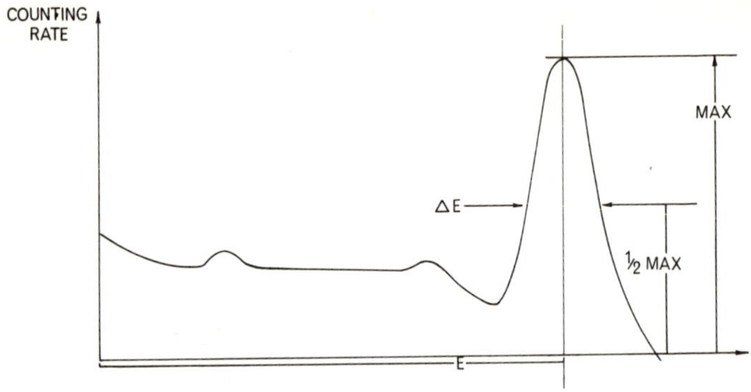

Fig.11. Gamma absorption spectrum of a hypothetical intermediate energy monoenergetic emitter.

discriminate Gaussian peaks of neighboring energies is heavily dependent on each detector's characteristics. This ability is usually called "resolution" and is usually evaluated as the net ratio of the full width at half maximum (ΔE) to the average energy of a given Gaussian peak (E) (see Fig.11). The "photofraction" is the ratio of the number of initially monoenergetic photons, counted within the full energy peak to those counted in the whole spectrum. The "full energy absolute efficiency" is the fraction of monoenergetic photons emitted from the source which is recorded within the full energy peak. It is equal to the product of the total efficiency times the photofraction.

The main X and gamma-ray detectors can be classified into the following groups: (*1*) gas filled detectors; (*2*) scintillation detectors; (*3*) semiconductor detectors.

GAS-FILLED DETECTORS

All the gas-filled detectors consist basically of a chamber containing two electrodes and filled with a suitable insulating gas. The absorption of any ionizing radiation by the gas gives rise to a number of ion pairs. If a d.c. voltage is applied between the electrodes, the ions are accelerated by the electrostatic field and collected at the electrodes where they produce an output pulse.

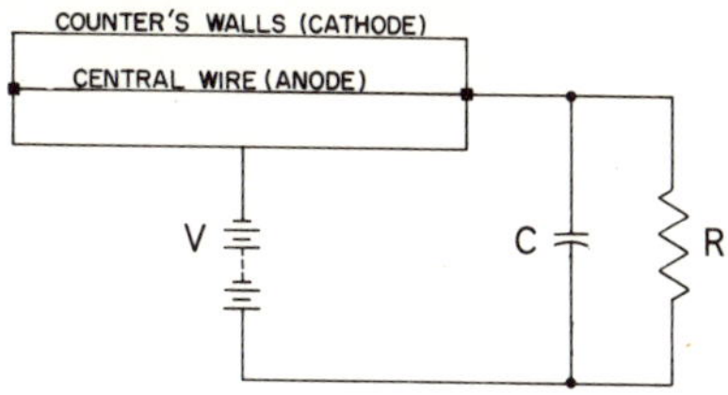

Fig.12. Schematic diagram of a typical gas filled detector.

Let us consider a typical gas-filled detector (Fig.12), i.e., a cylindrical metal chamber with a very thin central wire well insulated from the chamber walls. A voltage V is applied between the chamber walls and the central collecting wire by means of a resistance R shunted by a capacitor C. The central wire is positively

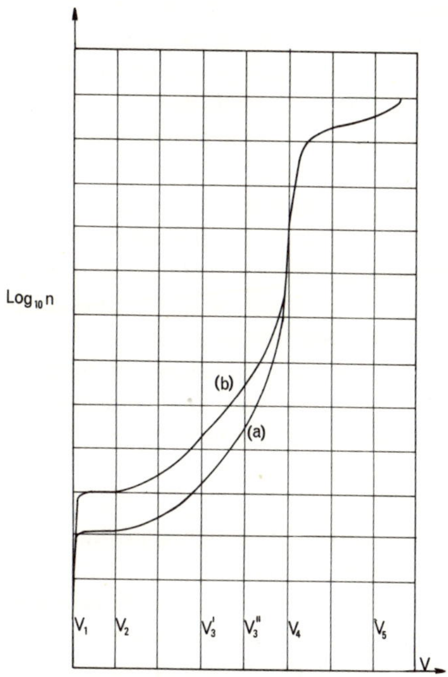

Fig.13. Semilogarithmic plot of the number of electrons collected at the central wire against the applied voltage. Curves a and b are relative to two ionizing radiations producing n_1 and n_2 ion pairs ($n_1 < n_2$) in the counter. (After WILKINSON, 1950; courtesy of University Press, Cambridge.)

charged relative to the chamber walls. Fig.13 is the classical plot by WILKINSON (1950) of the variation of the logarithm of the number of electrons collected at the central wire (log n) as a function of the applied voltage V for two ionizing radiations producing n_1 (curve a) and n_2 (curve b) ion pairs respectively ($n_1 < n_2$). When V is very low the collected electrons are less than the produced ones because some of them recombine with positive ions before reaching the wire. Starting from a value V_1, the acceleration that electrons undergo in the electrostatic field is sufficient to drive practically all of them to the central wire before recombination occurs. A further increase of V does not produce any sensitive change of the number of collected electrons up to a voltage V_2, where electrons acquire enough energy to produce secondary ionization. Each electron of a given energy produces the same number of secondary electrons and forms an avalanche; the number of electrons collected at the wire is greater than that of electrons produced by the primary radiation. The magnitude of the avalanches increases with increasing V but, as the single avalanches do not interfere with each other, the number of collected electrons is still proportional to the initial ionization; curves a and b in Fig.13 are thus essentially parallel between V_2–V'_3. With increasing V, a value is reached where the size of the avalanches increases to such an extent that their interaction becomes significant. At this value, the magnitude of avalanches is no longer proportional to the initial ionization and this value is of course lower for the most ionizing radiation (V'_3). From this point the essentially parallel curves begin to converge until they join at V_4. From this voltage, V_4, the number of collected electrons no longer depends on the initial ionization. The very rapid rise of the curves and their union indicates that the avalanches propagate throughout the length of the counter, one avalanche breeding one more and so on until, because of the different mobility of electrons and positive ions, a positive space charge forms close to the central wire reducing the field so that avalanches are smaller and cannot breed. This condition goes on up to V_5; from this value up, the discharge propagates indefinitely.

Three types of detectors have been developed working in three different regions of the diagram of Fig.13:

(*1*) Ionizing chambers, working in the range V_1–V_2. They are characterized by a practically complete collection of the produced ions without any gas amplification. The size of the output pulse is dependent on the initial ionization.

(*2*) Proportional counters, working in the range V_2–V_3. Gas amplification occurs, i.e., the number of collected electrons is greater than that produced in the range V_1–V_2. The size of the output pulse is still dependent, however, on the initial ionization.

(*3*) Geiger–Mueller (GM) counters, working in the range V_4–V_5. An extensive gas amplification occurs, the size of the output pulse being no more dependent on the initial ionization.

The actual values of the voltage defining each field depend of course on the characteristics of the counters, such as their size, nature and pressure of the filling gas, etc.

Geiger–Mueller (GM) counters

Before the availability of large, single scintillation crystals, most of the gamma radiometric work on rocks was carried out by means of Geiger–Mueller counters. These counters have been largely displaced by scintillation crystals which have higher intrinsic efficiencies and permit spectrometric analysis without differential absorbers. Geiger–Mueller counters are still useful because of some particular characteristics, such as the comparatively large size of the output pulses (not requiring any preamplification), the large variety of sizes and shapes, and the low cost.

The ordinary configuration of GM counters is the described cylindrical one (Fig.12). Any metal can be used for the chamber walls, and tubes made of glass with the inside surface covered by a conductive coating or tubes with external cathode (MAZE, 1946) have been built. The central wire is usually made of tungsten. GM tubes have been designed in a variety of shapes, including nearly 4π geometry spherical models (BROWN and EVANS, 1939; RAJEWSKI, 1943; SALVINI, 1948; JACOBS, 1964, 1965). A long external cathode flexible tube has been designed by RICHTER and GILLESPIE (1961), and RICHTER and BALLARD (1961). It is built by joining together short segments of conductive plastic materials (vinyl chloride polymers have been used) around which a bare copper wire is spiraled. The central wire is slipped inside a hose or sleeve and is maintained in an axial position for whatsoever disposition of the tube by means of polyethylene spacer disks or by a flexible skeleton of disks and rods. It can be built of any length and can be wrapped, coiled, and twisted into any nonlinear configuration. It works as a series of independent gas flow GM tubes because the discharge does not propagate through the polyethylene disks. Although it can be useful for measurements in caves and holes, etc., a serious drawback is the very low efficiency for gamma rays due to the nature of the walls.

The filling gas ordinarily used inside GM counters is argon, helium, or neon. Where a pure gas is used, the main discharge is followed by other discharges triggered from the electrons released because of the impact of positive ions on the cathode. These subsequent discharges can be quenched by an external circuit, but the most usual procedure is to add a small percentage of a gas having a lower ionization potential than the filling gas. Thus the ions of the main gas largely interact with the smaller ionization potential quenching gas rather than ejecting electrons from the cathode. Organic quenching gases were at first used; they undergo a rapid, quite extensive, decomposition with the consequence of restricting the life of the counter. The use of halogen quenching gases prolongs the life of the counter and allows one to apply higher voltages with negligible influence on the counter's life. Further advantages of halogen quenched counters are their steadier operation and background. Gas pressures inside GM counters are ordinarily less than one atmosphere, although higher pressures have been used in particular cases.

The counting rate, n, of GM counters shows a characteristic trend as a function of the applied voltage V (Fig. 14). The counting rate increases rapidly up to a value V_T (threshold voltage), above which there is a region in which the applied voltage does not affect the counting rate very much. This steady region is called the "plateau" and is usually 150–200 V across. The actual value of V_T is, of course, dependent on the characteristics of the tube. The operating voltage is chosen so as to be in the middle part of the plateau. It is in most cases in the range 800–1,500 V, but GM tubes with much lower operating voltages have been designed, such as, for instance, the bromine quenched tube described by GIMENEZ and LABEYRIE (1952), which operates at 350 V.

After the main discharge is produced, the electrical field around the wire falls below normal because of the presence of the positive ion sheath. Consequently if a further ionization occurs before the voltage of the wire is increased to a proper value, no discharge will be triggered. The range of time during which the counter is insensitive to the passage of any ionizing radiation is called the "dead time." Ordinarily in GM tubes it is of the order of 0.1–0.2 msec; therefore if a sensitive amplifier is used, the dead time of the whole system is close to that of the GM tube. A correction must be made for the unrecorded pulses. If τ is the dead time of the GM

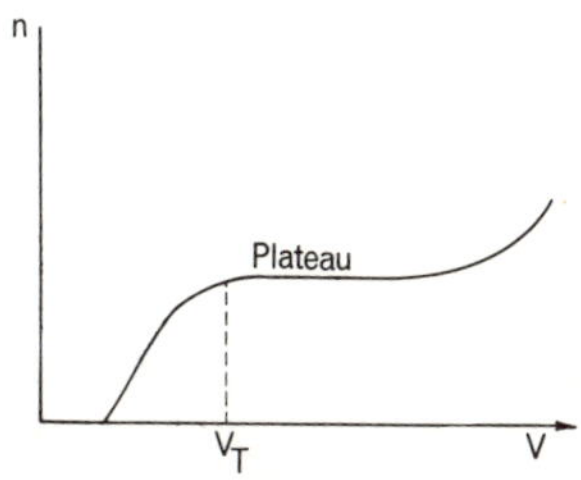

Fig.14. Qualitative variation of the counting rate against the applied voltage in a GM tube.

tube and n_r the pulses recorded for unit time, the pulses, n, which would have been recorded if the dead time was negligible, are:

$$n = \frac{n_r}{1 - n_r\tau} \tag{2.1}$$

Eq.(2.1) has been derived on the assumption that τ is independent of the counting rate, which is correct for comparatively low counting rates. It must be emphasized that the dead time usually given by the manufacturer is in most cases an instrumental dead time, i.e., it has been evaluated upon electric measurements and not upon measurement of radioactivity. As a consequence, it is based on the assumption of the arrival of ionizing radiations exactly spaced in time, which is not the case in radioactivity measurements. An extensive discussion of the dead time of GM counters has been made by STEVER (1942) and, more recently, ANDRESEN (1965) has discussed the problem.

The response of GM counters to gamma rays depends primarily

on the absorbing properties of the chamber walls, and on their opacity to electrons. The intrinsic efficiency is therefore mainly determined from the thickness and the nature of the material forming the walls, as well as from the energy of photons. Curves of the intrinsic efficiency as a function of the photon energy for brass, aluminium, and bismuth walls are reported in Fig. 15, after BRADT et al. (1946). The much higher intrinsic efficiency of bismuth walled tubes at low energies is due to the higher photoelectric cross section

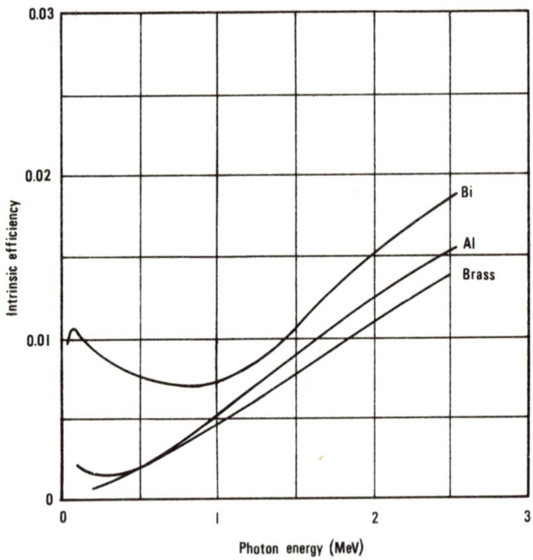

Fig. 15. Intrinsic efficiency of Bi, Al and brass cathode GM tubes. (After BRADT et al., 1946.)

of high atomic number substances. The nearly linear increase of all three curves in the range of Compton domain is a consequence of the nearly constant Compton cross section and of the concomitant increase of the secondary electrons' energy with increasing photon energy. Further curves have been reported by NORLING (1940); MEIER–LEIBNITZ (1946); JURNEY and MAIENSCHEIN (1949); HART et al. (1951) and SUZUKI and YUHARA (1961). DANCHENKO and MITROFANOV (1964) have designed a chlorine quenched GM tube where the use of a low atomic number material (beryllium) with thicknesses of the order of the range of low energy scattered electrons (about 0.1 mm) yielded an essentially energy independent gamma-ray efficiency in the range of prevailing Compton absorption (0.4–1.3 MeV).

A characteristic of GM tubes is that the background increases after periods of intense activity or after heating of the tube. The former effect is more sensitive with organic quenched than with halogen quenched tubes. A break in the operation is usually sufficient for the background to return to a normal value. The behavior of the background of a previously excited halogen quenched tube as a function of the temperature has been discussed by LESINSKI (1965). It increases up to a maximum around 150°C and then returns quickly to normal at higher temperatures. The amplitude of the maximum is directly dependent on the extent of the previous excitation. This phenomenon has been interpreted as due to the presence of electron trapping centers on the cathode surface, from which electrons escape when they acquire sufficient energy. The process goes on as long as the trapping centers are depleted of electrons. Ordinary GM tubes are designed for operation at temperatures not higher than 50–80°C. Special models are commercially available working at much higher temperatures. A description of a GM tube capable of operating up to 180°C has been reported by DRAGHICESCU (1961).

Proportional counters

In the late forties the work of KIRKWOOD et al. (1948) and CURRAN et al. (1948) have demonstrated the applicability of proportional counters to the spectrometry of X-rays and low energy gamma-rays. The main advantages of proportional counters in respect to GM tubes are the dependence of the output pulse size on the initial ionization, the shorter dead time allowing one to handle counting rates up to 10^6 counts/sec and the low background. They require a more complex electronic circuitry than do GM tubes. Compared to scintillation devices (Fig.20), they exhibit a much better resolving power for low energy photons. Semiconductors are the only competitive devices in this range of energies. But, even in respect to semiconductors, proportional counters have a better geometry because of the possibility of using internal sources. They can be made very transparent to high energy photons, further reducing their possible interferences on the low energy spectrum.

Proportional counters usually have cylindrical shapes. Some typical models are described in the review articles by CURRAN (1955, 1958), GATROUSIS et al. (1961) and CURRAN and WILSON (1965). Proportional counters especially designed for X-ray spectrometry have been described by many authors, including BERTOLINI et al. (1954); AMREHN and KULENKAMPF (1955); MILLER (1955); TROST (1958); GOGANOV and GOGOLEV (1963); LUKIRSKII et al. (1963); NARBUTT et al. (1963); MÜLLER (1964); BOWYER-STUART (1965); KOWNACKI and MARCINKOWSKI (1965); CULHANE et al. (1966), and LEGER et al. (1967).

Various materials can be used for the construction of proportional counters. The cathode is generally made of metal (steel, copper, brass, aluminium) or also of glass. The occurrence of radioactive impurities and the excitation of fluorescence X-rays by cosmic rays produce an undesirable background. GATROUSIS et al. (1961) have designed a low background counter having the cathode of graphite impregnated phenolic resins, coated with an alcoholic suspension of colloidal graphite, which has a high transmission for low energy photons. The central wire is usually of stainless steel, an essential requirement being the uniformity of the cross section through all its length; variations of the wire diameter have been reported to give rise to fluctuations of about 1% in the multiplication factor (WEST, 1953).

Where the photons are not collimated onto the middle part of the wire, a complication is introduced because the abnormal electric field at the ends of the wire affects the gas multiplication in these regions. An obvious way to reduce this "end effect" is to use counters which are long compared to their radius. An alternative method has been applied by COCKROFT and CURRAN (1951). It consists of adjusting the field at the ends of the wire by auxiliary electrodes. The "end effect" can also be corrected by dividing the wire into two unequal parts by a glass bead and by subtracting the recordings obtained from one part to those from the other part (ANGUS et al., 1949).

The most common filling gas is a mixture of 90% argon and 10% methane (usually referred as P 10 gas). Methane can be replaced with nitrogen, carbon dioxide, or other gases. Electronegative gases must be avoided because their tendency to capture electrons to form negative ions is a cause of wide fluctuations of the multiplication factor and affects the resolving power of the detector. Argon can be replaced by krypton or xenon. Fig.16 reports the variation of the percentage photoelectric and Compton absorption in 5 cm of the three gases at STP with the energy of photons. Argon appears to be preferable in X-ray spectrometry, whereas xenon and krypton have a better photoelectric efficiency in the range of low energetic gamma rays or high energy X-rays. After a photon has

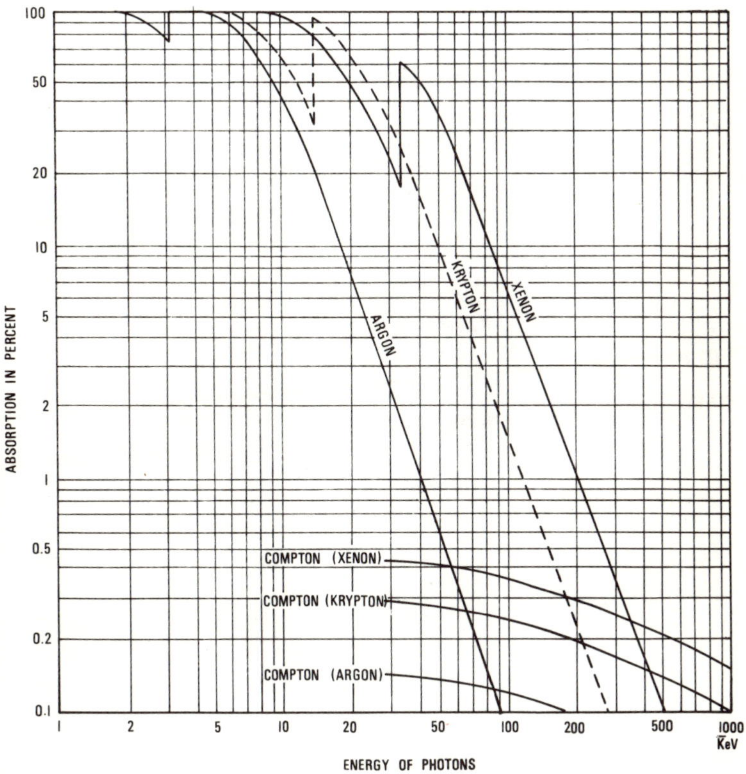

Fig.16. Variation of the fraction of the photoelectrically and Compton absorbed photons in 5 cm of Ar, Kr and Xe at STP against the photon energies. (After WEST, 1953.)

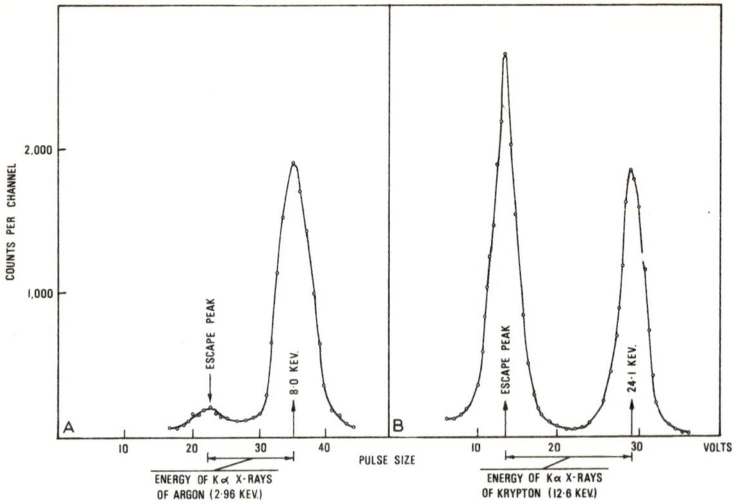

Fig.17A. K X-rays from ^{65}Zn in an Ar filled proportional counter. (Diameter 7.2 cm; filling gas: Ar = 37.7 cm, Methane = 4.1 cm.)
B. K X-rays from ^{113}Sn in a Kr filled proportional counter. (Diameter 5 cm; filling gas: Kr = 13.8 cm, Methane = 0.8 cm.) (After WEST, 1953.)

been photoelectrically absorbed by the gas, a K or L X-ray can be emitted by the excited gas atom (see Chapter 1, pp.11–12. According to whether the X-ray is or is not absorbed in the counter, a pulse height corresponding to the full energy of the initial photon or to the full energy minus the energy of the escaping X-ray will be recorded. As a consequence two peaks will be recorded: the full energy peak and the K X-ray escape peak (the L X-ray escape peak is usually not resolvable). The fluorescence yields of argon, krypton, and xenon are respectively 0.11 (±0.01), 0.67 (±0.03), and 0.81 (±0.05), according to WEST and ROTHWELL (1950). The consequent variation of the relative height of the K escape peak in an Ar and in a Kr filled proportional counter is shown in Fig.17. The relative size of the escape peak can be reduced by increasing the size of the counter and/or the pressure of the gas.

Fig.18 reports the quantum counting efficiencies[1] for different proportional counters and for a scintillation and a GM detector

[1] The quantum counting efficiency is defined as the fraction of photons incident on the front surface of the counter which expend some energy in the counter.

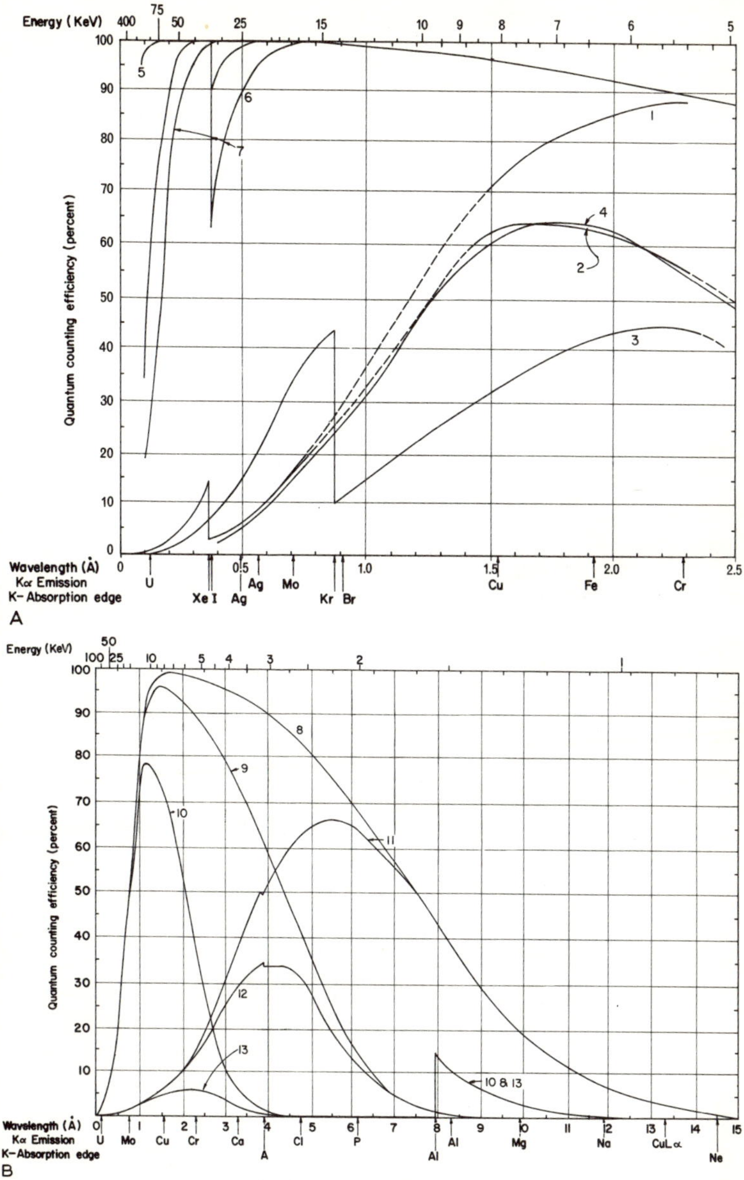

Fig.18. Quantum counting efficiencies of different X-ray detectors (see Table II) in the energy range 1–400 keV. (After TAYLOR and PARRISH, 1955.)

TABLE II

CHARACTERISTICS OF THE X-RAY DETECTORS USED FOR FIG.18

Curve no.	*Counter type*	*Filling gas*	*Pressure (torr)*	*Active length (cm)*	*Window thickness (cm)*
1	Proportional, side window	Xe	30	2.7	Be (0.0127)
2	Proportional, side window	Xe	30	2.7	Mica (0.0013) + Be (0.0127)
3	Proportional, side window	Kr	50	2.7	Mica (0.0013) + Be (0.0127)
4	Geiger–Mueller, end window	Ar	55	10.0	Mica (0.0013) + Be (0.0127)
5	Scintillation, NaI(Tl)			0.70	Be (0.0127)
6	Scintillation, NaI(Tl)			0.10	Be (0.0127)
7	Scintillation, NaI(Tl)			0.05	Be (0.0127)
8	Proportional, side window	Ar	76	2.7	Be (0.0025)
9	Proportional, side window	Ar	76	2.7	Be (0.0127)
10	Proportional, side window	Ar	76	2.7	Al (0.0025)
11	Proportional, side window	Ne 75.7 + Ar 0.3	76	2.7	Be (0.0025)
12	Proportional, side window	Ne 75.7 + Ar 0.3	76	2.7	Be (0.0127)
13	Proportional, side window	Ne 75.7 + Ar 0.3	76	2.7	Al (0.0025)

(see Table II), after TAYLOR and PARRISH (1955). BYRNE (1965) has evaluated theoretically the efficiency of a long cylindrical counter for X-rays emitted from a point source placed centrally on the walls of the counter.

The efficiency of a proportional counter is independent of the counting rate up to 1,500–2,000 counts/min. For higher counting rates the dead time effects become quite noticeable and the efficiency begins to decrease. The usual method for verifying the linearity of the response of a counter to different counting rates is to interpose a steadily increasing number of identical metal foils between a collimated monoenergetic source and the window of the counter and to plot the logarithm of the recorded counting rate against the total thickness of absorber (multifoil method). SHORT (1960) has suggested a method where only one absorbing foil is required and the counting rate is steadily decreased by varying the window width.

The backscattering of photons by the wall of the counter and the escape of primary photons and of secondary electrons from the sensitive volume are further disturbing effects. The former is mainly important when collimated external sources are measured. Low energy photons can in fact be backscattered without losing enough energy to be resolved from the photoelectric peak, causing therefore an overestimation of this peak. Since the Compton backscattering cross section decreases with increasing energy of photons (see Fig.10), this effect is more important when X-rays and low energy gamma rays are dealt with. Some instances of the interference of backscattered photons with photoelectric peaks are reported in Fig.19 after FINK and ROBINSON (1956).

The decrease of efficiency due to the escape of primary photons or of secondary electrons can be reduced by increasing gas pressure or the size of the counter. Additional methods to reduce this effect are the use of two coaxial multiwire cathodes defining two anticoincidence coaxial counters (DREVER et al., 1957). Anticoincidence counters of this type have been described by MANDUCHI and ZANNONI (1962) and by SIEVERT and SCHMIDT (1966). The escape of the electrons can be reduced by the use of a magnetic field which constrains the electrons to follow spiral trajectories (CURRAN et al., 1950).

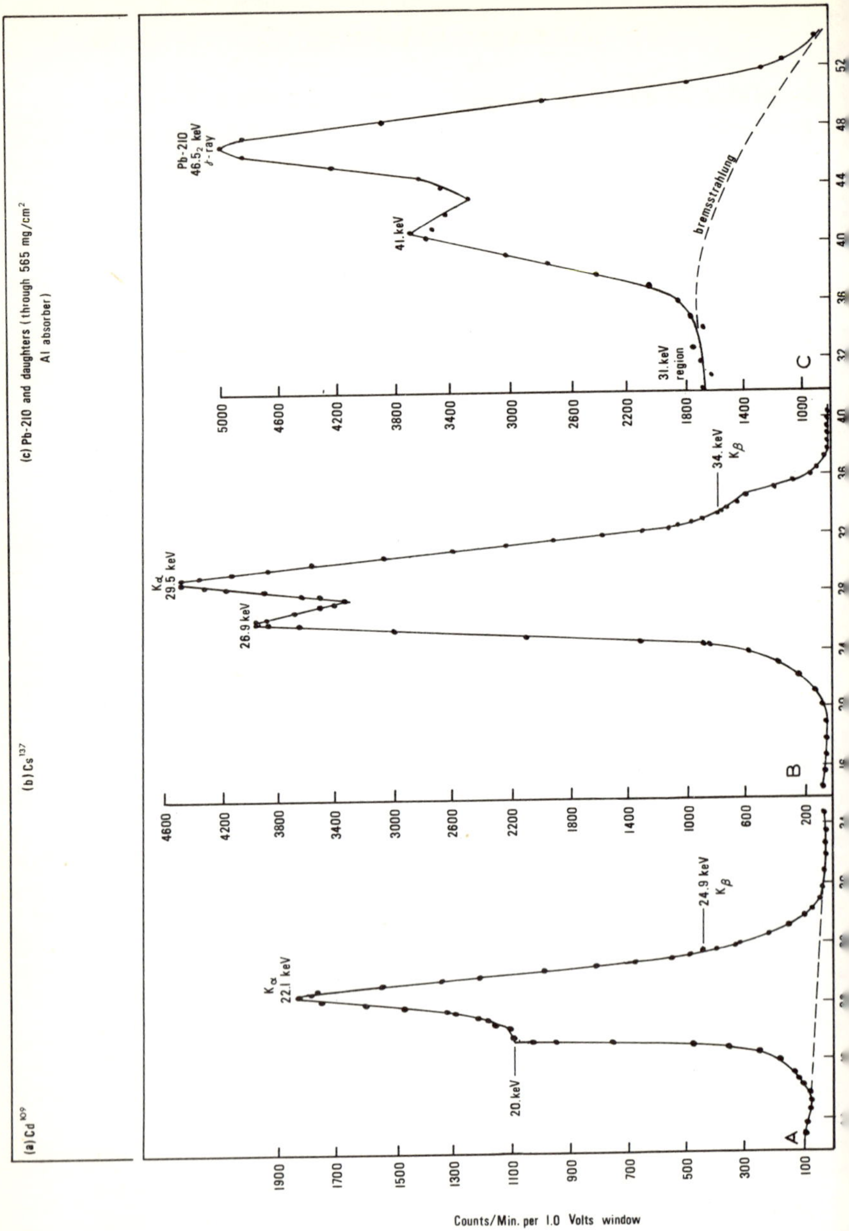

Fig.19. Examples of interference of the photons backscattered from the counter's walls on the photoelectric peak, showing also the energy dependence of the effect. All three external sources have been identically collimated and the X-rays measured by a proportional counter with brass, aluminum lined, walls and filled with P-10 gas at 2.1 atm. A. ^{109}Cd. B. ^{137}Cs. C. ^{210}Pb and daughters (through 565 mg/cm^2 Al absorber). (After FINK and ROBINSON, 1956.)

As pointed out at the beginning, a most outstanding feature of proportional counters is their satisfactory resolution. Fig.20, comparing the energy spectra obtained by a NaI(Tl) crystal and by a

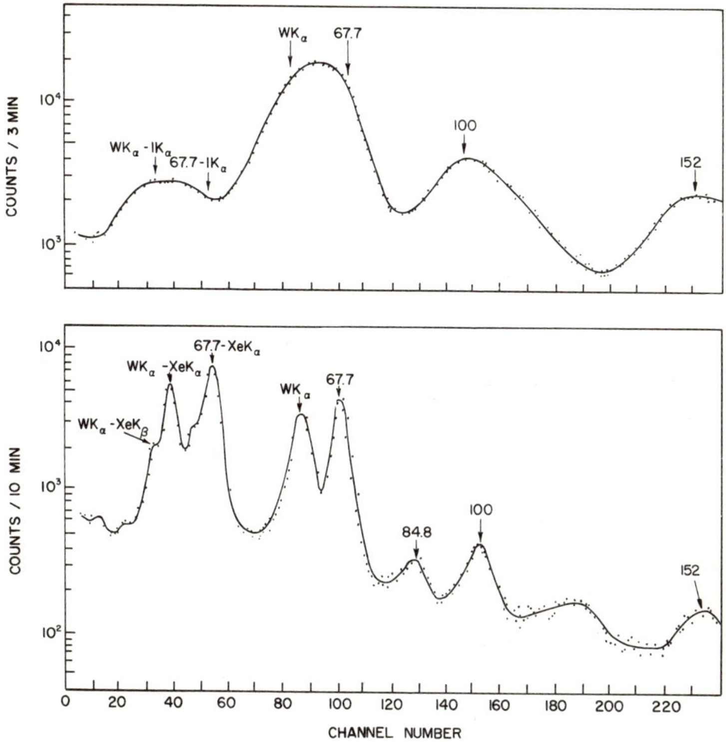

Fig.20. Comparison of a 3 × 3 inch NaI(Tl) scintillation spectrum (upper spectrum) and a proportional counter spectrum of a ^{182}Ta source in the 42–160 keV range. The proportional counter is 1 mm Al walled, with a 11.4 cm long and 5 cm diameter active section, and filled with Xe at 2 atm. (After HANSEN et al., 1965; courtesy of North-Holland Publ. Co.)

proportional counter in the energy range 40–180 keV, requires no further comments. Where the fluctuations due to the associated electronic circuitry are not considered, the resolution depends on the spread of the output pulse height. This is mostly affected by the fluctuations of the number of formed ion pairs and of the gas multiplication factor. The variance (d_p) of the output pulse size is therefore given by:

$$d_p = d_0 p_A^2 + d_A p_0 \tag{2.2}$$

d_0 and p_0 being the variance and the average number of ions formed by primary ionization in the counter and d_A and p_A the corresponding quantities for the number of ions produced in an avalanche by a single ionization electron.

FANO (1947) gives for d_0 the range:

$$p_0/2 \leq d_0 \leq p_0/3 \tag{2.3}$$

The fluctuation of the gas multiplication (p_A) depends on several factors. ROSE and KORFF (1941) have derived theoretically a formula relating the gas multiplication factor p_A to the applied voltage (V_0), the threshold voltage (V_t), the capacity for unit length ($\frac{1}{2}C$), the radius of the wire (r_1), the number of molecules of filling gas per volume unit (N) and the rate of change (K) of the ionization cross section for energies above the ionization potential:

$$p_A = \exp 2(KNCr_1 V_0)^{\frac{1}{2}}[(V_0/V_t)^{\frac{1}{2}} - 1] \tag{2.4}$$

All the factors of eq.(2.4) can be directly measured; V_t can be obtained by extrapolating the nearly linear region of the curve of $\log p_A$ versus V to $p_A = 1$. Eq.(2.4) has been derived upon the assumptions that the photoelectric effect at the cathode, recombinations and electron attachments to neutral molecules, and fluctuations of the energy loss and of the specific ionization can all be neglected. CURRAN and CRAGGS (1949) have slightly modified eq.(2.4) by introducing an expression relating V_t to some characteristics of the counter and of the filling gas. KHRISTOV (1957) and DIETHORN (1956) have derived similar equations based upon slightly different assumptions. Disagreement between these formulae and the experimental data are often reported. KOCHAROV and KOROLEV (1963), on the contrary, report satisfactory agreement between experimental data and all the above reported equations. A new equation has been recently derived by ZASTAWNY (1966) and experimentally verified with satisfactory results for argon, nitrogen and carbon dioxide filled proportional counters (ZASTAWNY and MIZERACZYK, 1966; ZASTAWNY, 1967).

Fluctuations of all the quantities of eq.(2.4) will result of course in a spread of the output pulse height. An experimental determination of d_A has given (CURRAN et al., 1949):

$$d_A = 0.68 p_A^2 \tag{2.5}$$

From eq.(2.2), (2.3), and (2.5) the relative variance $d_p/(p_0 p_A)^2$ of the pulse height is:

$$\frac{d_p}{(p_0 p_A)^2} = \frac{1}{\alpha p_0} + \frac{0.68}{p_0} \tag{2.6}$$

where α is a number between 2 and 3.

End effect and ionization losses due to electron attracting impurities are further sources of fluctuations of the output pulse height. Also a high gamma background has been reported to affect the multiplication factor (TOLCHENOV, 1960).

High pressure ionization chambers

Ionization chambers are mostly applied to the detection of very ionizing particles, e.g., alphas or protons. Their application to the detection of electrons released from the absorption of photons requires very high pressures in order that the produced number of ion pairs is sufficient to produce a detectable output pulse. Pressures of the order of 100 atmospheres have been used.

Although high pressure ionization chambers have a sensitivity allowing, when a 4π geometry is used, a fairly accurate determination of activities as low as 0.01 μ Ci/cm^3 (COHEN et al., 1964), they do not offer substantial advantages in respect to the previously described detectors for the radiometry of rocks. They have been mainly used for dosimetry and relative calibration purposes. Their characteristics and uses have been discussed by ROSSI and STAUB (1949) and by WILSON et al. (1950). Recently developed designs have been described by COHEN et al. (1964), HENGELCKE and HUBNER (1965), BLANC et al. (1966), ERDELYVARI and FEHER (1967).

COLLIE et al. (1950) have used a high pressure ion chamber filled with deuterium for studying the photodisintegration of deuterons by gamma photons of 2.62 MeV from ^{208}Tl (see Chapter 1, pp.21–22) with the detection of the protons separated from neutrons by the photodisintegration. A chamber of this type could also have an inverse application and be used for the determination of the 2.62 MeV activity from ^{208}Tl. With a threshold energy of 2.23 MeV for the $^{2}H(\gamma, n)^{1}H$ reaction, only a very small interference would occur from the minor 2.2 MeV gamma from bismuth-214 of the uranium-238 series.

SCINTILLATION DETECTORS

Since the pioneer investigations of radioactivity beginning in 1896 some substances (called "phosphors") were known to have the property of emitting light flashes when struck by heavily ionizing particles. Where the counting rates were not too high, the individual light flashes could be visually observed and counted under a microscope. These early scintillation detecting assemblies were called "spintharoscopes." This simple but tedious counting technique was made obsolete with the introduction of the gas filled detectors in the 1920's. In the middle 1940's Curran and Baker first substituted an electronic device, the multiplier phototube, for the human eye. Since that time increasing research effort was devoted to scintillation detectors and Hofstadter in 1948 showed that sodium iodide thallium activated, NaI(Tl), crystals were detectors of great utility for gamma rays. These crystals had a widespread application for both the total and spectrometric determinations of gamma-rays. HURLEY (1954) first applied them to the gamma radiometry of rocks, and most of the recent data about the concentrations of natural radioisotopes in natural systems are from gamma-ray scintillation spectrometry.

A present day scintillation detector consists schematically of two parts: a phosphor and a multiplier phototube producing an output pulse whose height is proportional to the intensity of the light flash (scintillation) produced in the phosphor. The scintillation detector's noteworthy characteristics are the proportionality between output pulse height and energy expended in the phosphor by an incident photon, the satisfactory cross section that most of them have for both photoelectric and Compton absorption of photons with energies up to a few MeV, its very fast response to ionizing radiations (the decay time of the light output is of the order of microseconds for inorganic phosphors), its compactness, versatility, and simplicity of use. The scintillation detector's main shortcoming is the poor resolution as compared to proportional counters and semiconductors. Details concerning the design and construction of scintillation counters can be found in many excellent monographs, e.g. those by BIRKS (1953, 1964a); BELL (1955); MOTT and SUTTON (1958); CROUTHAMEL (1960); MURRAY (1962); NEILER and BELL (1965), and KAISER (1966). The mechanism of the scintillation process has been discussed by CURRAN (1953); MURRAY (1962); BIRKS (1964a, b); REHMANN and SCHARMANN (1966).

Let us consider a gamma photon of energy E passing through the crystal and expending in it all or a part of its energy. The energy expended by the photon in the phosphor is transformed into a voltage output pulse according to the following successive steps: (*1*) the photon gives an amount E_1 of energy ($E_1 \leq E$) to a charged particle, usually an electron; (*2*) the electron dissipates a fraction AE_1 of its energy ($A \leq 1$) in the phosphor; (*3*) the energy AE_1 is converted by the phosphor, with a conversion efficiency Y ($Y = pE_p/AE_1$), into p light photons, each having an average energy E_p; (*4*) a fraction G of the emitted photons goes toward the photocathode of the photomultiplier. Moreover, because of the imperfect transparency J ($J < 1$) of the crystal to its own fluorescence radiation, a fraction of them will be absorbed by the phosphor itself; (*5*) the photons reaching the photocathode are converted into photoelectrons with an efficiency P (P = number of photoelectrons per photon).

In conclusion, the total number of photoelectrons (n) produced at the photocathode as a consequence of the absorption of an energy E from a gamma ray is:

$$n = \frac{E_1 AYJGP}{E_p} \tag{2.7}$$

The photoelectrons are then multiplied by the multiplier phototube and, if U is the overall amplification of the tube, the total charge Q at the final collector plate of the phototube is:

$$Q = neU \tag{2.8}$$

TABLE III

SOME PROPERTIES OF THE MAIN GAMMA-RAY SCINTILLATORS

Scintillator	*Density* (g/cm^3)	*Z*	*Decay time* (μsec)	*Conversion efficiency*	*Emission spectrum*	*References*
NaI(Tl)	3.67	Na 11 I 53	0.3	13%	Band centered at 4,200 Å	Birks (1953); Murray (1962)
CsI(Tl)	4.51	Cs 55 I 53	1.1	~6%	from blue to red with maximum in red	Birks (1953); Murray (1962)
CsI(Na)	4.51	Cs 55 I 53	0.65	~12%	band centered between 4,000 and 5,000 Å	Brinckmann (1965)
CsF	3.59	Cs 55 F 9	0.005	~0.7%	band centered at 3,900 Å	Murray (1962)
CaI_2(Eu)	3.96	Ca 20 I 53	0.79	twice than NaI(Tl)	peak at 4,700 Å	Hofstadter et al. (1964); Aliaga-Kelly and Nicoll (1966)
CaI_2	3.96	Ca 20 I 53	0.55	twice than NaI(Tl)	peak at 4,100 Å	Hofstadter et al. (1964); Aliaga-Kelly and Nicoll (1966)
$CdWO_4$	7.9	Cd 48 W 74 O 8	6	~8%	peak at 5,300 Å	Birks (1953); Mott and Sutton (1958)
Plastic	1.2	H 1 C 6	$\sim 10^{-3}$	~2%	band centered between 3,800 and 4,500 Å	Murray (1962); Mott and Sutton (1958)

where e is the electronic charge. The proportionality between Q and E_1 allows the scintillation counter to be used as a spectrometer.

Table III summarizes the characteristics of the best known gamma-ray scintillators. NaI(Tl) is by far the most commonly used. Its comparatively high density and the presence of high atomic number iodine atoms ($Z = 53$) give rise to large photoelectric, Compton and pair production absorption cross sections. Fig.21 shows the NaI(Tl) absorption coefficients, after GRODSTEIN (1957). The decay time of the light output is adequate for normal counting purposes. Being hygroscopic, NaI(Tl) crystals must be maintained in sealed containers; exposure to moisture produces the formation of a yellowish absorbing surface giving rise to a large drop in resolution. Therefore NaI(Tl) crystals are usually wrapped in an aluminum can about 1 mm thick, whose inner side is coated with a

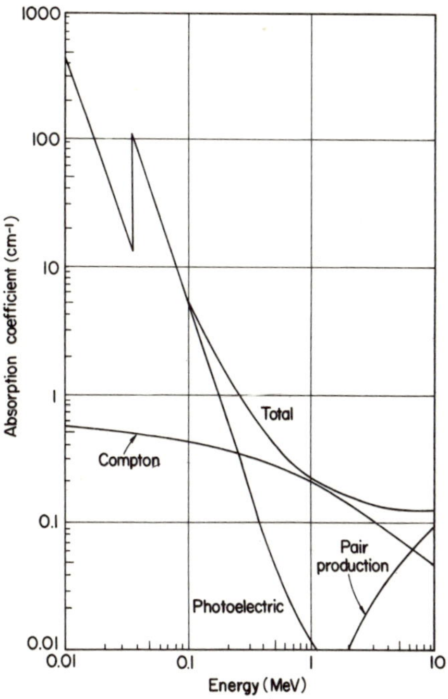

Fig.21. Linear absorption coefficients of NaI. (After the data reported by GRODSTEIN, 1957.)

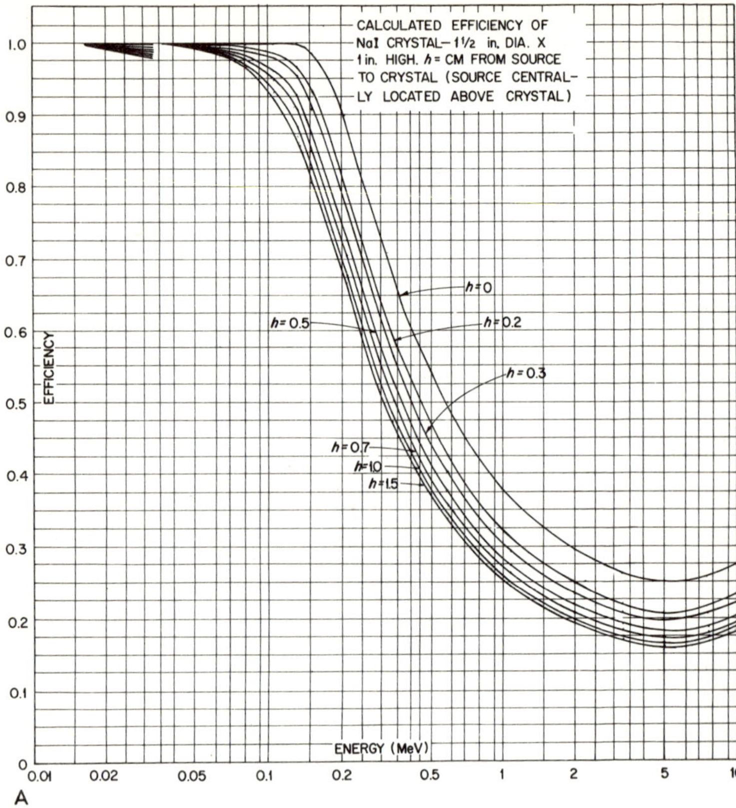

Fig.22. Calculated intrinsic efficiency curves of $1\frac{1}{2} \times 1$ inch NaI(Tl) crystals for various crystal–source distances. (After NEILER and BELL, 1965; courtesy of North-Holland Publ. Co.)

magnesia or alumina layer which reflects the light photons produced within the phosphor (see Fig.30). The conversion efficiency (Y) is one of the highest and the crystals are highly transparent to their own fluorescence radiation (peaking around 4,200 Å). This characteristic allows the use of very large crystals. Fig.22 shows the intrinsic efficiency curves of cylindrical $1\frac{1}{2}$-inch diameter by 1 inch high NaI(Tl) crystals for point sources located at various distances along the vertical axis of the crystal. The intrinsic efficiency decreases with increasing distance from $h = 0$ to

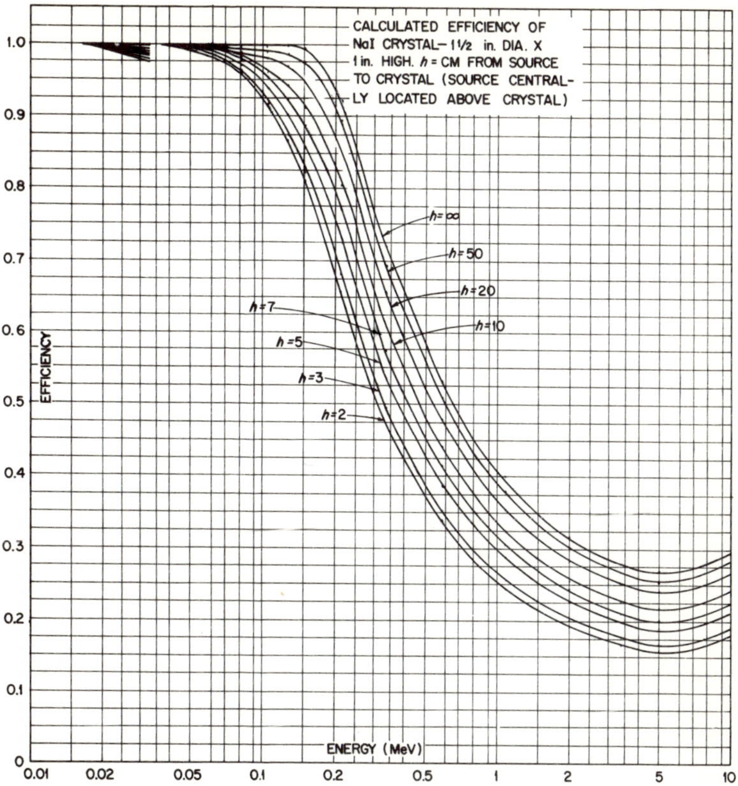

B

$h = 1.5$ cm (Fig.22*a*) and increases with distance from $h = 2.0$ cm to $h = \infty$ (Fig.22*b*). This is due to the so-called "edge effect"; gamma-rays from a source very close or very distant from the detector have a high probability of encountering a nearly constant thickness, whereas gamma-rays from intermediate distance have a higher chance of passing through the edges of the crystal, crossing therefore a smaller thickness. CROUTHAMEL (1960) has tabulated total efficiencies for cylindrical NaI(Tl) crystals of different sizes for various source distances. ZERBY and MORAN (1961) and JARCZYK et al. (1962) have tabulated photofractions. The geometry is greatly improved in well-type crystals, where the source is located in a well drilled along the vertical axis of the cylinder. Total efficiencies for two Harshaw well-type NaI(Tl) crystals for cylindrical and disk sources are shown in Fig. 23 based upon the data reported by SNYDER (1967). Further efficiency data for well-type crystals are reported from HABERER (1965); MISHRA et al. (1965); SNYDER and GYOREY (1965); and TIWARI and KONDAIAH (1966). Photofractions for well-type crystals are reported by SNYDER and KNOLL (1966).

In dealing with spectrometric determinations of a source having several gamma emitters, the easiest way to discriminate the intensity of a given gamma radiation is to count only the pulses recorded in the full energy peak; it is therefore important that the

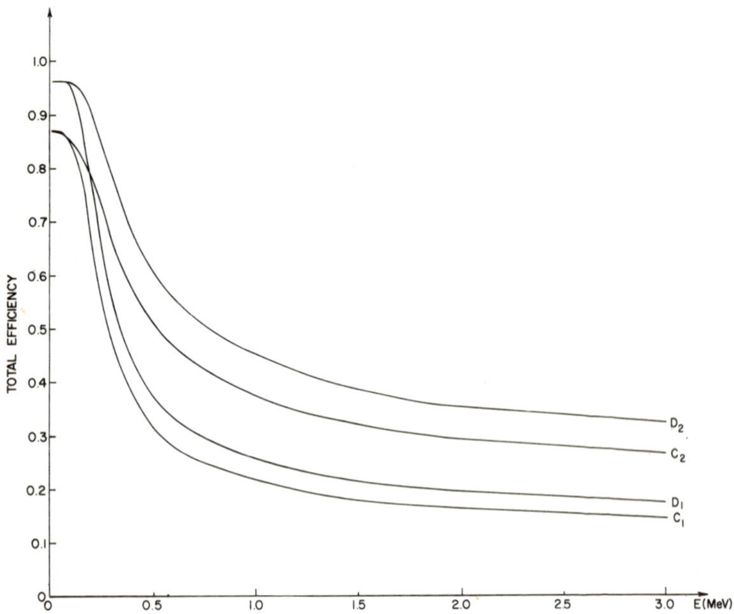

Fig.23. Total efficiency curves of two well-type crystals for cylindrical (*C*) and disk (*D*) isotropic sources. Self absorption not considered. Curves with subscript *1* are relative to Harshaw 8F8 2 × 2 inch (5.08 × 5.08 cm) NaI(Tl) crystal and curves with subscript *2* to Harshaw 12 AW(12)-W2 3 × 3 inch (7.62 × 7.62 cm) crystal. The dimensions of the wells are: height, 3.81 cm and diameter, 2.857 cm, for both crystals. Cylindric source: 3.50 cm high, top of the source at level of the top of the crystal; radius: 1.429 cm. Disk source: 1.43 cm diameter, 0.50 cm distant from the well bottom. (Plotted from the data by SNYDER, 1967.)

photofraction is as high as possible. The photons contributing to the full energy peak are not only those photoelectrically absorbed but, since the time interval between multiple interactions by a single photon is much shorter than the decay time of the phosphor light output, also those expending their energy in the crystal by multiple Compton interaction ending with a photoelectric absorption. The most direct way to increase the photofraction is therefore to increase the size of the crystal so that the probability of the latter event is enhanced. The effect of the crystal size on the photofraction is clearly shown in Fig.24. Crystals up to 13.5 inch (40.6 cm) have been grown. Large crystals of this order, usually referred to as "total absorption crystals", are also used as anticoincidence annuli. They have generally a poorer resolution, mostly because of non-uniform transmission of the visible photons produced in the crystals and they use several multiplier phototubes.

Iodine X-ray escape is a significant event up to about 100 keV in crystals $1\frac{1}{2} \times 1$ inch in size (see Fig.25). At higher energies, gamma-rays penetrate deeper in the crystal, and furthermore,

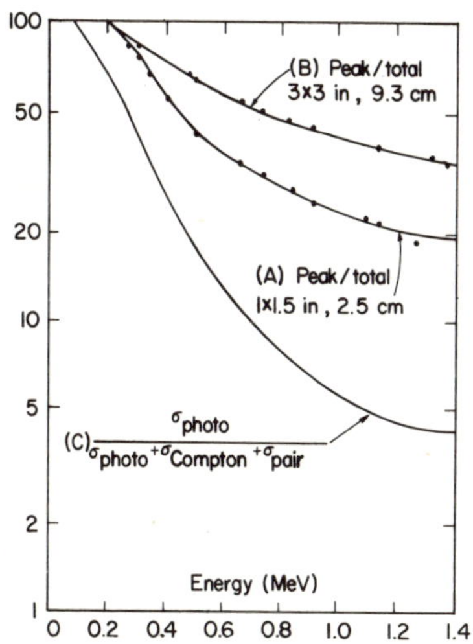

Fig.24. Photofraction vs. photon energy for $1 \times 1\frac{1}{2}$ inch (*A*), and 3×3 inch (*B*) NaI(Tl) crystals. Curve *C* is the ratio of the photoelectric to total absorption cross sections for NaI. (After NEILER and BELL, 1965; courtesy of North-Holland Publ. Co.)

since the iodine escape X-ray is only of 28 keV energy, the escape peak is poorly resolved from the full energy peaks at high energies.

Eq.(2.7) shows that factors inherent to the crystal affecting the number of electrons collected at the photocathode are the conversion efficiency (Y), the transparency (J) to its own fluorescence radiation, the fraction of energy (AE_1) dissipated by the electrons and the fraction (G) of photons reaching the photocathode. Fluctuations of each of these parameters will result in a spread of n, and will therefore affect the resolution. The resolution therefore is primarily dependent on the presence of inhomogeneities, variations of the light reflections, and imperfections of the crystal. Because the number of secondary photons produced in a given crystal is directly proportional to the energy the incident photon has given off in the crystal, the statistical variation is higher the lower is the energy spent in the phosphor. Fig.26 shows the variation of the resolution with the photon energy, after the data reported by KELLY et al. (1956). It is common practice to refer the resolution of a gamma-ray scintillator to the 0.661 MeV gamma-ray from ^{137}Cs. Resolutions with ^{137}Cs as low as 6.0% have been reported; however, the resolution of acceptable commercial crystals is in the range 7–8%. Well type crystals have resolutions generally higher than 10%. With smaller crystals the resolution is higher because of the more uniform transmission of light; small sized crystals are most fitted to the detection of X-rays and low energy gamma-rays because they are transparent to high energy radiations. Compton interference with low energy peaks is therefore reduced. NaI(Tl) slices of the thickness of 0.25 inch or less are commercially available.

NaI(Tl) crystals always contain some K impurity (usually of the order of some p.p.m.), which is particularly troublesome because of the natural radioactive isotope ^{40}K. ^{40}K emits gamma-rays with 1.46 MeV of energy (see Chapter 4, and Appendix 2), producing therefore a background which is mostly undesirable when determinations are made on low activity samples. The magnitude of the ^{40}K background is, of course, dependent on both the K content and the volume of the crystal. MANAGAN and CROUTHAMEL (1960) have estimated a background of about 30 pulses per minute in the energy range 25–1,580 keV due to a 2 p.p.m. content of ^{40}K in a 7 inch diameter by $3\frac{1}{2}$ inch high NaI(Tl) crystal. Considerable

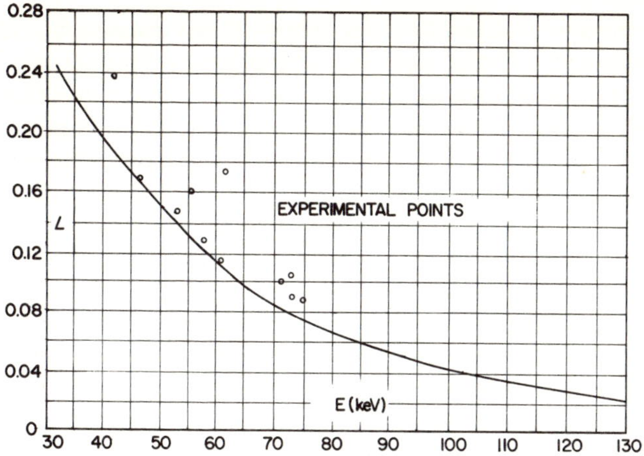

Fig.25. Iodine X-ray escape from NaI with incoming radiation normal to the surface and some experimental values. *L* is the ratio of the number of X-rays leaving the crystal to that of gamma rays entering a semi-infinite crystal normal to its surface. (After NEILER and BELL, 1965; courtesy of North-Holland Publ. Co.)

effort has been devoted in recent years to growing crystals with low potassium content; concentrations lower than 1 p.p.m. have been obtained so far (ALIAGA-KELLY and NICOLL, 1966).

CsI(Tl) crystals are the main alternative to NaI(Tl). They are not hygroscopic and, being softer and more plastic than NaI(Tl), can better withstand shocks and vibrations, as well as sudden

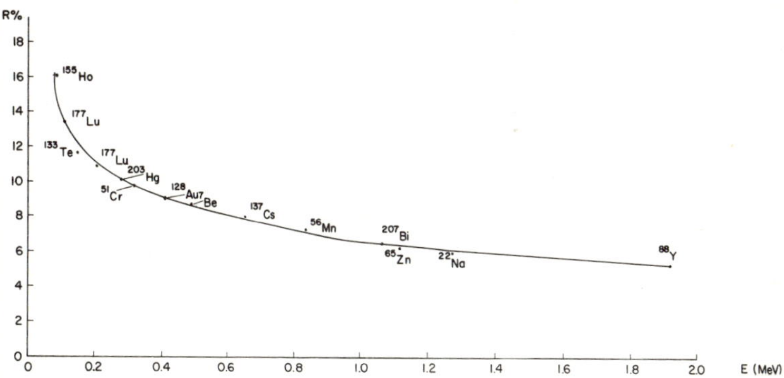

Fig.26. Resolution (*R*) vs. photon energy for selected NaI(Tl) crystal–photomultiplier combinations. (Plotted from the data by KELLY et al., 1956.)

temperature changes or large temperature excursions. These features make CsI(Tl) crystals competitive to NaI(Tl) for field work, bore hole logging, airborne surveys or space research. In Fig.27 are given the variation of the photoelectric, Compton, pair production, and total absorption coefficients versus the photon energy. The presence of Cs ($Z = 55$) instead of Na causes the absorption coefficients to be slightly higher than the corresponding ones for NaI(Tl). The photofraction of CsI(Tl), 2 × 2 inch, well-type crystal is 1.5–2.4 times that of an equivalent NaI(Tl) crystal for energies in the range 0.4–3.0 MeV (see Fig.28). The higher photofraction of CsI(Tl) is, however, obtained at the expense of a slightly poorer resolution (MILLER and SNOW, 1960*a*). A positive advantage of all CsI(Tl) crystals is that the ^{40}K content is usually much lower than is the case with NaI(Tl).

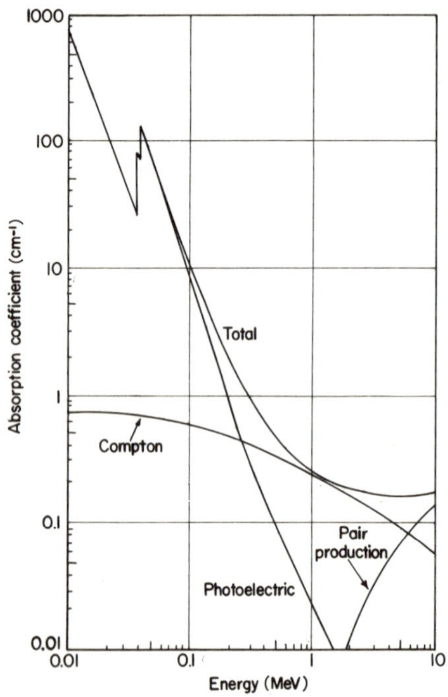

Fig.27. Linear absorption coefficients of CsI. (After Harshaw, scintillation phosphors; courtesy of Harshaw Chemical Co.)

BRINCKMANN (1965) has reported that CsI(Na) crystals have better scintillation characteristics than CsI(Tl); the light output decay constant is of the same order of magnitude as NaI(Tl) and the conversion efficiency is twice that of CsI(Tl), giving therefore rise to a better resolution. A further interesting feature is that the maximum conversion efficiency is at about 80°C (KAISER, 1966), so that it is particularly fitted for measurements in environments with slightly above normal temperatures.

HOFSTADTER et al. (1964) have shown that CaI_2 and CaI_2(Eu) crystals have an exceptional conversion efficiency, resulting in a resolution as high as 5.2% for 0.661 MeV gamma-rays from ^{137}Cs. The photofraction is slightly higher than NaI(Tl) (see Fig.28). A restriction for the use of such crystals is, however, that attempts to grow crystals larger than 0.25 inch (6.4 mm) have failed to date because of the formation of parallel cleavage cracks.

$CaWO_4$ crystals are seldom used in spite of their high density and atomic number resulting in a high gamma detection efficiency. A major disadvantage is their high index of refraction, having as a consequence a poor light coupling.

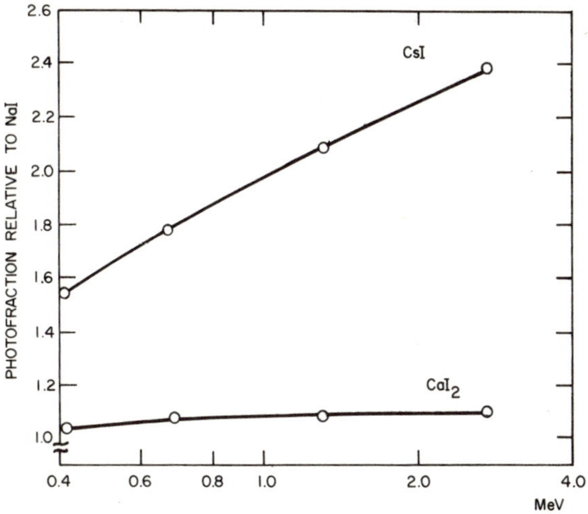

Fig.28. Ratios of photofractions of 2 × 2 inch (5.1 × 5.1 cm), well type CsI and CaI_2 crystals to NaI crystals of the same size against the photon energy. The size of the well is: 2.857 cm diameter by 3.81 cm height. (After SNYDER and KNOLL, 1966; courtesy of North-Holland Publ. Co.)

Plastic scintillators are made by dissolving an organic scintillator in an organic solvent. Their very low density and atomic number make them much less efficient than inorganic scintillators for the detection of gamma-rays. They can be made quite large and have been applied to total gamma-ray counting in airborne surveys. The ease of fabricating large sizes favors their use as anticoincidence annuli. Fig.29 shows a characteristic spectrum obtained by a plastic scintillator; it is a spectrum of ^{22}Na. The two full energy peaks that are expected at 0.511 and 1.28 MeV are not evident and the spectrum essentially consists of a continuum due wholly to Compton scattering.

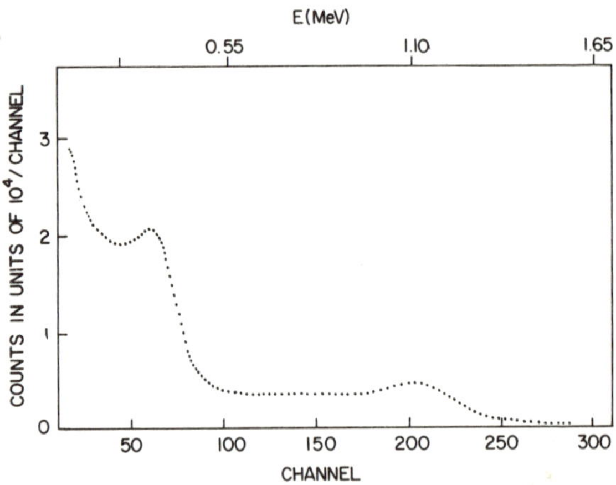

Fig.29. Gamma-ray spectrum of an ^{22}Na source by a plastic scintillator. (After GRISSOM and KOEHLER, 1965; courtesy of North-Holland Publ. Co.)

The resolution of any of the described scintillation assemblies is critically dependent on the multiplier phototube being matched to the phosphor. The factors U and P of Eq.(2.7) and (2.8) are in fact inherent to the multiplier phototube. In order to achieve optimum results, a photomultiplier must have a very fast linear response (about one nanosecond), high amplification (about 10^6), low noise, minimum spread in the collection time, and spectral response matching closely the spectral output of the scintillator. A number of commercially available multiplier phototubes, manufactured by many companies including R.C.A., DuMont, EMI, CBS, Philips, etc., meet such requirements. Usually the tubes have a flat end window, directly coupled to the phosphor by a high viscosity silicone oil. Dow Corning DC200 silicone oil is mostly used with NaI(Tl); ophtalmological petroleum jelly or Dow Corning C-2-0057, being gelatinous, are helpful in applications where the oil could flow out. Light guides or pipes are used where the end window is not flat or when the phosphor must be used at some distance from the phototube. The optical coupling of crystals is a most critical operation in that the resolution of the assembly is

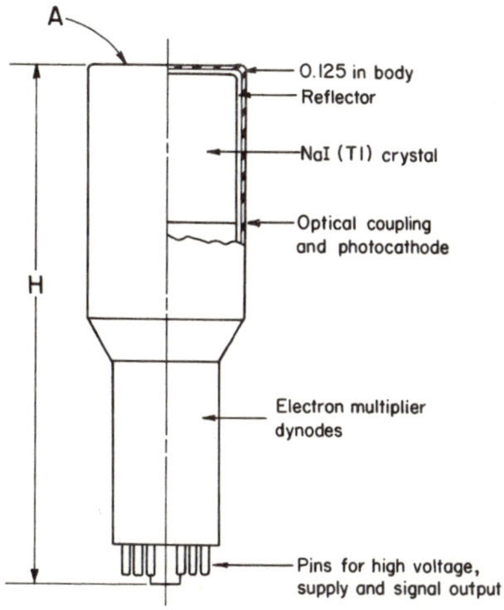

Fig.30. Integral line Harshaw type 12 S crystal (H = 5.78 inch (14.6 cm); crystal diameter 3 inch (7.6 cm); crystal height = from $\frac{1}{4}$ inch (0.67 cm) to 3 inch (7.6 cm) A = 0.019 (0.048) entrance window). (From Harshaw Catalog, 1968; courtesy of Harshaw Chemical Co.)

critically dependent on the coupling. Although the operation is not very complicated, it requires a certain skill to eliminate air bubbles, etc. In order to prevent any mishandling of this delicate part of the detecting assembly, many manufacturers have made sealed scintillators–multiplier phototube assemblies (integral lines), which are more readily usable and more rugged. Fig.30 illustrates a typical integral line assembly. One should note in Fig.30 that the integral assembly eliminates one transparent window; hence only the epoxy resin used for the optical coupling is between the crystal and the envelope of the multiplier phototube.

In order that an optimum photocathode efficiency and a minimum spread of P (Eq.2.7) are achieved, the photocathode must have high efficiency where the phosphor emission peaks and must be as uniform as possible. A photocathode generally consists of a Cs–Sb semitransparent layer on the interior surface of the glass (or silica) end window. Recently, multi-alkali cathodes have been developed having a higher efficiency in the region where the NaI(Tl) emission peaks and a higher stability.

The photoelectrons emitted from the photocathode are multiplied by secondary emission from a series (9–14) of electrodes (dynodes), between which an increasing voltage is applied. U (Eq.2.8) depends on the emission characteristics of the dynodes and on the applied voltage. Consequently these quantities must be kept as constant as possible in order that the variance of U is minimized. Dynodes are usually made of Cs–Sb or Ag–Mg. A dependence of the output pulse height on the counting rate has been observed for some multiplier phototubes. This effect has been ascribed to the "dynode fatigue", resulting in a change of the Cs content of the dynode surface. Even with Ag–Mg dynodes, photocathodes are formed from Cs vapor and therefore Cs will be present on the dynode surface. Recently, Cu–Be dynodes have been used with mixed results. The performance of a multiplier phototube is affected by the presence of quite small magnetic fields, even by the earth's magnetic field. Magnetic fields deflect the moving electrons and hence can produce a decrease of the overall gain. Such effects can be minimized by surrounding the multiplier phototube with a cylindrical mu metal shield.

A strong temperature dependent effect of multiplier phototubes is the thermo-ionic emission of electrons at the photocathode (dark current), which can give rise to noise of the order of 5,000 electrons/cm^2 sec at room temperature. The following equation applies:

$$n_e = zT^2 \exp(-b/T) \tag{2.9}$$

where n_e is the emission rate of thermo-ionic electrons, z the photocathode area, b a function of the photocathode material, and T the absolute temperature. An effective means of dealing with n_e is to cool and thermostat the photocathode.

Natural radioisotope impurities (mainly ^{40}K) in the construction materials of the multiplier phototubes contribute, of course, to the background. A background reduction has been obtained by the use of fused silica faced tubes or by all ceramic and metal tubes. A cheaper approach may be the use of light pipes to absorb a part of the radiations from the phototube itself.

SEMICONDUCTOR DETECTORS[1]

Semiconductor detectors are presently the most exciting alternative to scintillation counters. Their most striking feature is the resolution typically at least ten times higher than NaI(Tl), allowing a very detailed study of the fine structure of the most complex spectra. Further advantages are the fast pulse rise time (about one nanosecond), the compactness and, compared to gas filled detectors, the higher absorption coefficients. Their main short-coming is presently the small size of the sensitive volume, resulting in a low intrinsic efficiency for gamma photons of intermediate and high energy. This feature is particularly troublesome for the gamma spectrometrist concerned with natural systems, in which one often deals with Th and U concentrations of the order of p.p.m. or less. In order to improve the intrinsic efficiency, the main effort in recent years has been devoted to the increase of the sensitive volume. However, even with the present efficiency limitations, semiconductor detectors are very useful wherever a detailed study of a complex gamma or X-ray spectrum is needed, and the specific activity of the source is sufficiently high, e.g., for the determination of the neutron activation products. The requirement of liquid nitrogen cooling for the Ge(Li) detectors operates against their use in the field and in logging.

---

[1] These detectors are sometimes also referred to as "solid state detectors". As this designation is ambiguous in that it would seem to include all the detectors based upon the interaction of radiations or charged particles with solids (e.g., it should include also scintillation counters), the designation of "semiconductor counters" is preferred for the present work.

The principle of operation of such detectors is basically the collection of the electric charge released in a solid by the absorption of photons or charged particles. Van Heerden first counted β^- rays by applying a voltage across an insulator, a silver chloride crystal. The energy bands of the outer electrons in an insulator are schematically represented in Fig.31A; all the electrons are accommodated in the valence band (and of course in the deeper lying bands). The conduction band is usually empty and no movement of electric charge occurs. The valence and the conduction bands are

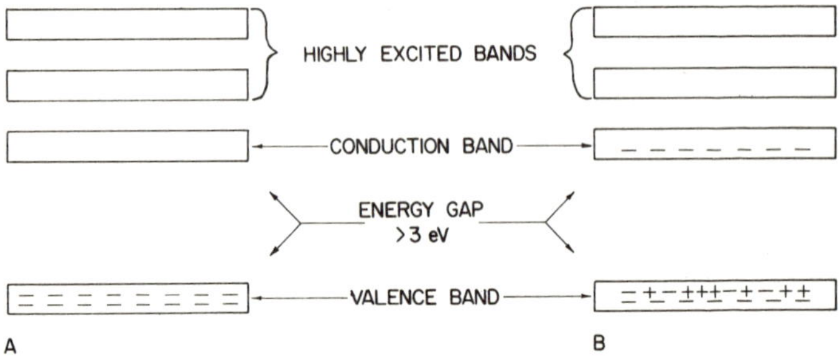

Fig.31. Schematic representation of the outer electronic bands of an insulator at rest (A) and after the absorption of an ionizing radiation (B).

separated by an energy gap usually wider than 3 eV. When an ionizing particle or radiation is absorbed by the insulator, it can transfer to a single electron enough energy to raise it from the valence band (or even from the deeper lying bands) to the highly excited bands; the removal of a negative charge from the valence band appears as a positive hole in the valence band. The highly excited states exist only about one picosecond after which the electrons are lowered to the conduction band, many more electron–hole pairs being formed in the process. The final situation is that of Fig.31B with positive holes at the top of the valence band and an equal number of electrons in the bottom of the conduction band. When two electrodes are placed on two parallel faces of the crystal and a voltage is applied, the charge carriers move to the appropriate electrodes and the induced charge can be detected.

In order to have good detection characteristics, the material of the detector should fulfill the following requirements as close as possible:

(*1*) It should contain very few free electric charge carriers (i.e., few electrons in the conduction band and/or holes in the valence band) in the operating condition; such carriers will be swept to the electrodes and will tend to conceal or to spread the desired signal.

(*2*) It should not contain a significant number of trapping centers capable of holding electrons or holes and therefore giving rise to an incomplete charge collection. Furthermore, trapped electrons and holes produce an electric field in the crystal which, being opposite to the applied field, causes the current to decrease with the time (polarization effect).

(*3*) Recombination of holes and electrons during the charge collection process must be very small.

(*4*) Charge collection must be as fast as possible; holes and electrons must therefore have a high mobility.

(*5*) In order to get a good resolution, the average energy to produce an electron–hole pair must be as low as possible. The average energy to produce an electron–hole pair is >3 eV in an insulator and <3 eV in a semiconductor. These values are very low as compared to the average energy required to form an electron–ion pair in an ionization chamber (about 29 eV) or to produce a photoelectron at the photocathode of a scintillation counter (about 500 eV). Consequently more electron–hole pairs are produced in a semiconductor or insulator by a radiation of a given energy, resulting in improved statistics and, therefore, in better resolution.

The failure to fulfill completely requirement (*2*) because of the frequent presence of impurities and lattice defects, is the main limitation in the use of the above described detectors. McKay first showed that the coupling of two different types of semiconductors (*p* and *n* types) gives rise to a device having better detection properties than homogeneous insulators or semiconductors. This coupling gives higher mobility of the charge carriers and a uniform electric field due to the uniform distribution of impurities and a very low trapping effect.

Fig.32A shows the energy band of outer electrons in a semiconductor. It differs from the energy bands in an insulator in two important features: (*1*) the energy gap between conduction and

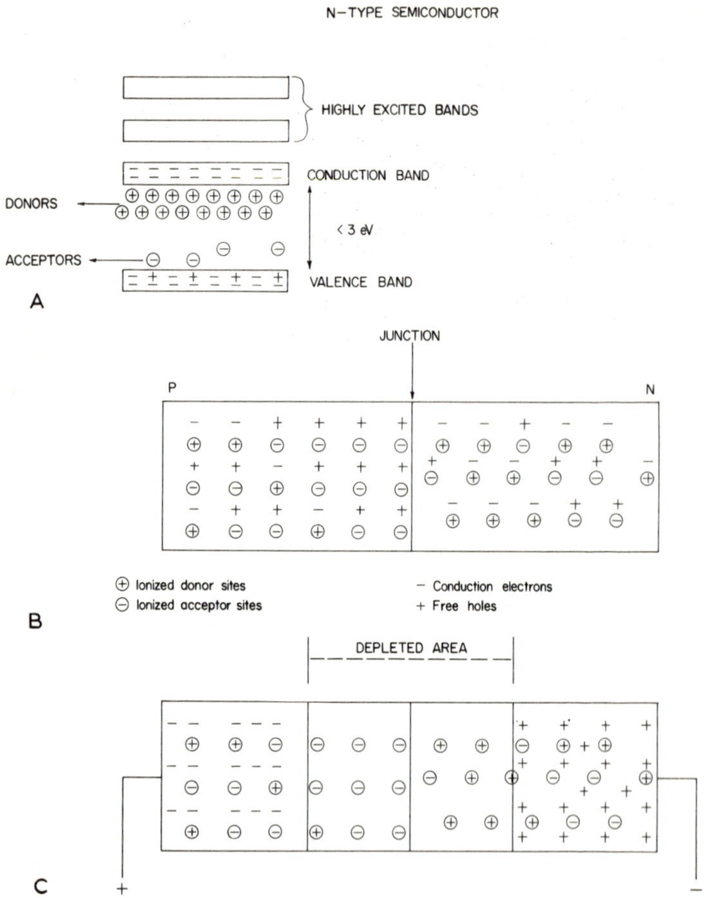

Fig.32. Schematic representation of the outer electronic bands of an *n*-type semiconductor (A) and of a *p–n* junction at rest (B) and after a reverse bias has been applied (C).

valence bands is usually lower than 3 eV; and (*2*) impurity energy states are present close to the crystal's energy bands. The impurities, the presence of which allows a semiconductor to conduct, can be schematically gathered into two types: (*1*) those giving electrons to the conduction band (donor impurities); and (*2*) those accepting electrons from the valence band (acceptor impurities). An "*n*-type" semiconductor has an excess of donor impurities (i.e., an excess of conduction electrons over free holes, such as in Fig.32A); a "*p*-type" semiconductor has an excess of acceptor impurities (i.e., an excess of free holes over conduction electrons). When an "*n*-type" is coupled to a "*p*-type" semiconductor (*p–n* junction), the distribution of free holes, conduction electrons, and impurity sites is that shown in Fig.32B. When a reverse bias is applied, all the conduction electrons and the free holes are swept to the electrodes and a "depletion area" is formed where a space charge due to the ionized impurity sites exists (Fig.32C). When the free charge carriers are all removed, no conduction of electricity is possible until new electron–hole pairs are formed. The depletion area is the sensitive area of a *p–n* junction detector.

A *p–n* junction can be formed by diffusing a donor impurity, e.g., phosphorus, into a *p*-type semiconductor, e.g., silicon (n^+p junction), or vice versa by diffusing an acceptor impurity, e.g. gallium or boron, into a *n*-type semiconductor (p^+n junction). Still another way is to form a surface *p*-type layer on a *n*-type semiconductor and to coat it with a thin, very pure metal layer, e.g., gold or silver (surface barrier detectors). These techniques only produce a small sensitive volume and, hence such detectors are inadequate for use in gamma-ray spectrometry. A method for increasing the depletion volume has been described by PELL (1960). The method is based upon the drifting of lithium (a donor impurity) ions into a semiconductor crystal under an electric field at temperatures in the range 100°–400°C. Usually *p*-type silicon or germanium with a uniform distribution of acceptor sites are employed. Detectors of this kind are called lithium drifted silicon, Si(Li), or germanium, Ge(Li), detectors. Details concerning the theory and the manufacturing of all the above mentioned types of semiconductors are reported by TAYLOR (1963), PRICE (1964), GIBSON et al. (1965), DEARNALEY and NORTHRUP (1966).

The technology of manufacture of pure Si and Ge ingots is highly developed because of the extensive application they have in the field of transistors; thus they are also currently employed as semiconductor detectors. Because of its high atomic number and density, Ge has a higher absorption coefficient (Fig.33) than does

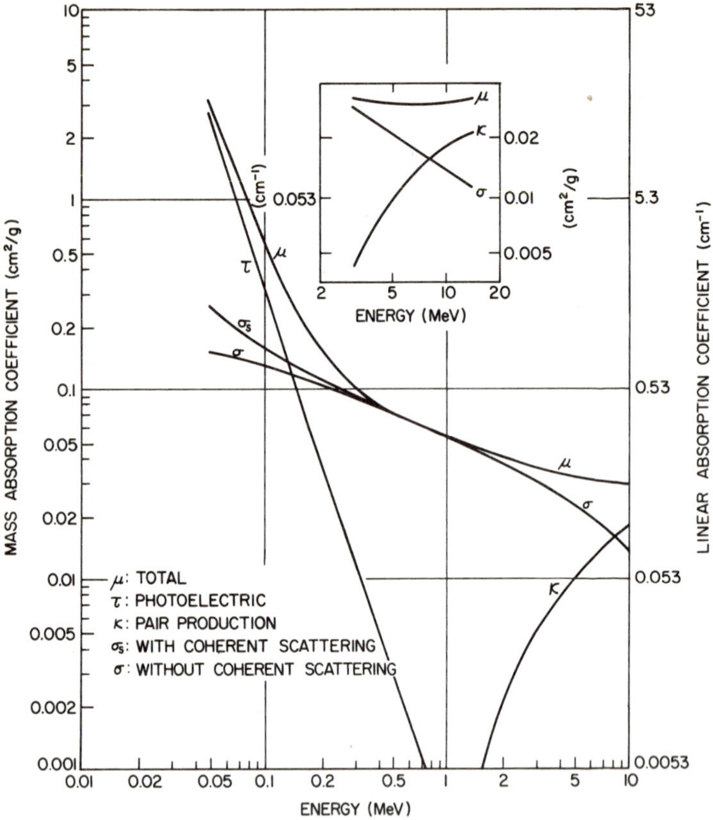

Fig.33. Mass and linear absorption coefficients of Ge. (After CHAPMAN, 1967; courtesy of North-Holland Publ. Co.)

Si (Fig.34). Table IV summarizes the main characteristics of both elements. Ge(Li) is much better fitted for the spectrometry of gamma-rays, whereas the use of Si(Li) detectors is restricted to intermediate and low energy X-ray spectrometry.

The application of both Ge(Li) and Si(Li) detectors to X-ray spectrometry has been discussed by BOWMAN et al. (1966). They report results obtained with a 2.5 mm thick depletion, Si(Li) and a Ge(Li), with a 3.0 mm thick depletion layer. The former has a 50% intrinsic efficiency at about 30 keV, whereas the Ge(Li)

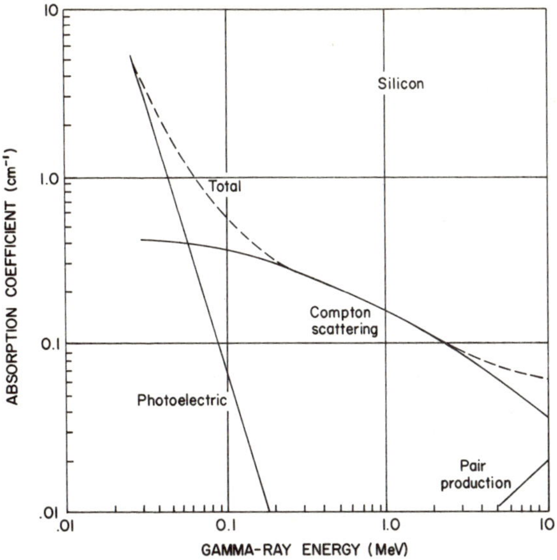

Fig.34. Linear absorption coefficients of Si. (After DEARNALEY and NORTHRUP, 1963; courtesy of Wiley and Sons.)

detector has the same intrinsic efficiency at 100 keV. Fig.35 shows the resolution obtained for K_{α_1} and K_{α_2} and K_{β_1} and K_{β_2} X-rays from two neighboring elements, Pt ($Z = 79$) and Au ($Z = 78$). Furthermore, all K_α and K_β X-rays from Au and Pb ($Z = 82$) are resolved. A low energy cutoff is set by self absorption in the sources, counter window absorption, and various sources of noise. The lower limit is about 4 keV, so that X-rays from elements below scandium ($Z = 21$) cannot be satisfactorily analyzed.

The resolution of such detectors can be fully exploited if they are matched to a field effect transistor (FET) preamplifier. Resolutions as high as 2 keV (half width at half maximum) for 1.33 MeV gamma-rays from ^{60}Co have been reported (STRAUSS et al., 1967).

The intrinsic efficiency for intermediate and high energy gamma-rays is comparatively low with Ge(Li). Efforts to increase the sensitive volume have led to the development of total depletion detectors (ZIEMBA et al., 1962) and of large sensitive volume Ge(Li) detectors, either by a new technique of Li drifting (TAVENDALE, 1965) or by

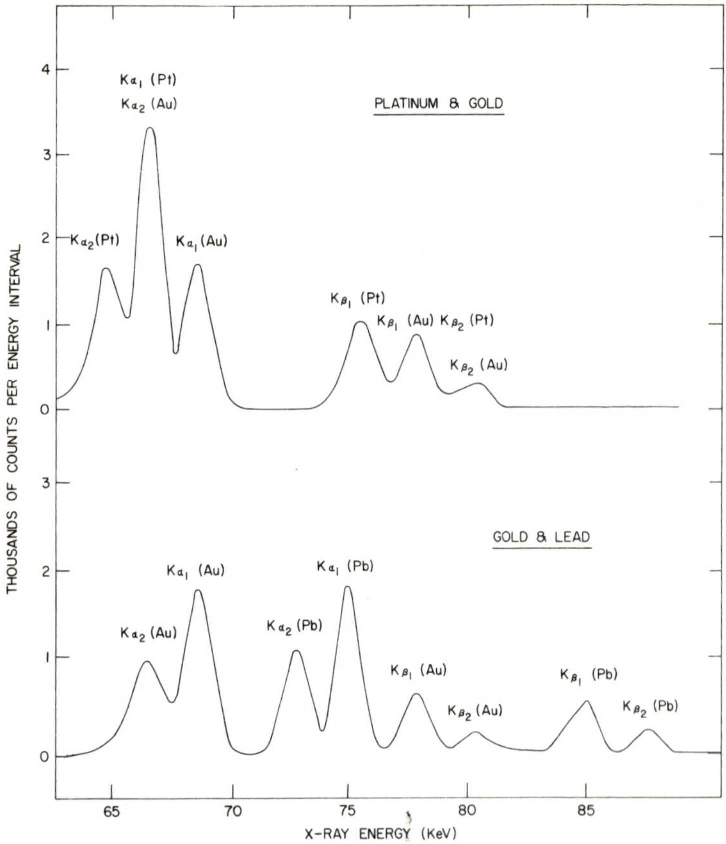

Fig.35. X-ray spectra from Pt–Au and Pb–Au metal foils obtained by a Ge (Li) detector. (After BOWMAN et al., 1966; copyright American Association for the Advancement of Science.)

the use of coaxial diodes (MALM, 1966; MALM and FOWLER, 1966; TAVENDALE, 1966). Fig.36 reports for different detector sizes the curves of intrinsic full energy peak efficiency (defined as the ratio of the gamma-rays recorded under the full energy peak to those of the same initial energy reaching the sensitive volume of the detector). It should be noted in Fig.36 that the largest Ge(Li) have an efficiency remarkably lower than ordinary NaI(Tl) crystals at

TABLE IV

SOME PROPERTIES OF SILICON AND GERMANIUM
(After GOULDING, 1966)

	Si	*Ge*
Atomic number	14	32
Atomic weight	28.06	72.60
Density (300°K)	2.33 g/cm^3	5.33 g/cm^3
Dielectric constant	12	16
Energy gap (300°K)	1.106 eV	0.67 eV
Energy gap (T°K)	$1.205–2.8 \cdot 10^{-4}$ eV	$0.782–3.4 \cdot 10^{-4}$ eV
Electron mobility (300°K)	1,350 cm^2/Vsec	3,900 cm^2/Vsec
Hole mobility (300°K)	480 cm^2/Vsec	1,900 cm^2/Vsec
Electron mobility (T°K)	$2.1 \cdot 10^9 T^{-2.5}$* cm^2/Vsec	$4.9 \cdot 10^7 T^{-1.66}$* cm^2/Vsec
Hole mobility (T°K)	$2.3 \cdot 10^9 T^{-2.7}$* cm^2/Vsec	$1.05 \cdot 10^9 T^{-2.33}$* cm^2/Vsec
Energy per hole–electron pair	3.66 eV	2.96 eV
Intrinsic carrier density (300°K)	$1.5 \cdot 10^{10}$/cm	$2.4 \cdot 10^{13}$/cm
Intrinsic carrier density (T°K)	$2.8 \cdot 10^{16} T^{\frac{3}{2}} e^{-6450/T}$/cm	$9.7 \cdot 10^{15} T^{\frac{3}{2}} e^{-4350/T}$/cm

* Measured only over a limited temperature range (about 100–300°K for Ge and 160–400°K for Si).

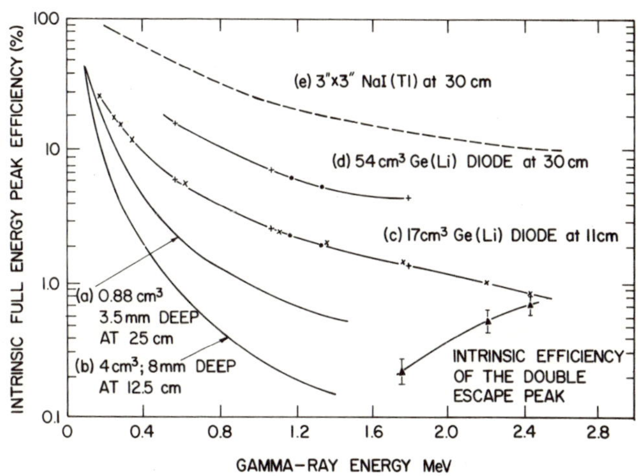

Fig.36. Intrinsic full energy peak efficiencies for different sized Ge(Li) detectors and for a 3 × 3 inch NaI(Tl) crystal. The distance of the source from the detector is also indicated. (After MALM and FOWLER, 1966.)

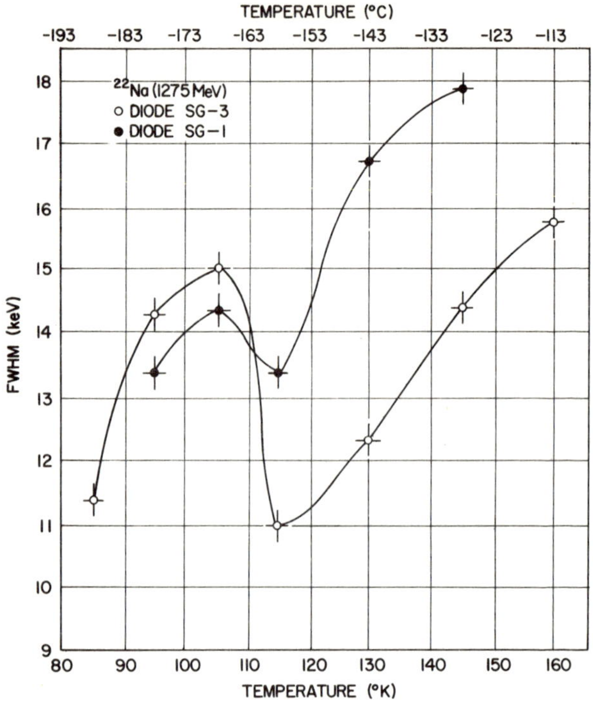

Fig.37. Variation of the resolution (expressed as full width at half maximum of the ^{22}Na 1.275 MeV full energy peak) as a function of the temperature for two 20 mm diameter Ge (Li) diodes with a 3.5 mm thick depletion area. (After EL SHISHINI and ZOBEL, 1966.)

present. An alternative approach to increasing the efficiency is to improve the geometry by using several Ge(Li) detectors working in parallel (LALOVIC et al., 1967; MALM, 1967), or by annular detectors such as that described by LEVY et al. (1967). The recent progress in semiconductor detector design and performance has been discussed by GLOS (1964); GOULDING (1964, 1966); SHIRLEY (1965); HOLLANDER (1966); MAYER (1966); MANN (1967).

Ge(Li) detectors must be operated at liquid nitrogen temperature (77°K) in order to reduce to very low values (less than one nanoampere) the leakage current due to thermal excitation of electrons and to increase the charge carriers' mobility. The temperature dependence of the resolution of Ge(Li) detectors is clearly shown in Fig.37. Furthermore, the leakage currents are very sensitive to surface conditions and therefore the detectors must be cooled under vacuum. In order to prevent redistribution and precipitation of Li, they must be stored at temperatures lower than 222°K. (−50°C).

When FET preamplifiers are used, an optimum performance is obtained if they are cooled and located close to the detector. It can therefore be convenient to locate both in the same enclosure. Vacuum failures and warming up accidents can have a very deleterious effect. The encapsulation of the detectors in hermetic seals (WEBB et al., 1966) is a means to minimize such accidents. Si(Li) detectors are less sensitive to temperature, but optimum performances are obtained when they are operated at very low temperatures.

CHAPTER 3

Laboratory and Field Instrumentation

LABORATORY INSTRUMENTATION

A laboratory for gamma radiometry of rocks should be designed after careful consideration of the end use of the data to be obtained. The main characteristics of such a laboratory should depend on the order of magnitude of the activities to be assayed. In a laboratory where high grade uranium or thorium ores are to be investigated, the main problem may be the dead time and the overloading of the electronic circuitry. Both detectors and electronic components must have a very fast response in order to avoid a high dead time correction. A rough shielding is generally adequate to obtain a satisfactory sample counting rate/background counting rate ratio.

If the laboratory must be devoted mainly to general geochemical studies of natural radioisotopes, a quite low sample counting rate is to be expected. The average abundances of the main radioelements in the earth's crust are: K = 2.1%; U = 3 p.p.m.; Th = 12 p.p.m. This means that a 100 g sample of average crustal materials emits only 6–7 gammas/sec due to ^{40}K and about 20–25 gammas/sec due to ^{238}U, ^{235}U, and ^{232}Th series in secular equilibrium. A hundred times lower contents are commonly observed in ultrabasic igneous rocks, and contents as low as picograms have been reported for iron meteorites.

The instrumentation for the measurement of low activities must fulfill the following requirements in order to get effective counting statistics in a practical time: (*1*) high intrinsic efficiency of detector; (*2*) good stability and minimum noise by all the electronic components; (*3*) low background.

The requirement of a high intrinsic efficiency makes inorganic scintillation crystals the most favorable detectors at present.

Among them, alkali halides, mainly NaI(Tl), being available in larger sizes are most used. Dealing with average crustal materials, 3 × 3 inch (76 × 76 mm) or 5 × 5 inch (126 × 126 mm) solid crystals are adequate, but larger sizes or improved geometry arrangements are advisable when lower contents are measured or when only a few grams of sample are available. In either case, solid crystals are generally to be preferred to well type crystals, not only for their higher resolution, but also because the poorer geometry they offer is more than compensated by the higher intrinsic efficiency and by the possibility of using much larger samples. Commercially available crystals have in fact usually 5 cm depth by 2.5 cm diameter wells; consequently only a few tens of grams can be measured. Solid crystals on the other hand offer the possibility of using containers completely surrounding the crystal, save the optical coupling side, with the consequent possibility of using even some kg of crushed rock uniformly distributed in a few cm thick layer. Furthermore, the cost of well type crystals is remarkably higher than that of analogous solid crystals. The geometry can be greatly improved by using two or more crystals viewing the same source. A very simple arrangement of this kind has been described by ADAMS (1964), consisting of two 3 × 3 inch NaI(Tl) integral line assemblies facing each other, and viewing therefore the opposite sides of an opaque container placed between them. The joint resolution is 8% for the ^{137}Cs full energy peak, i.e., still remarkably higher than that obtainable by well type crystals.

Ge(Li) and Si(Li) detectors with sensitive volumes large enough to have an intrinsic efficiency comparable with that of a 3 × 3 inch NaI(Tl) crystal have not yet been built. Therefore, although good results for the analysis of rocks have been reported (MATHEVON et al., 1967), there is no reason at present to prefer them to NaI(Tl), unless the requirement of a more detailed spectrum does not overshadow that of a higher sensitivity. Si(Li) and Ge(Li) detectors have a satisfactory intrinsic efficiency in the X-ray range, where they, as well as proportional counters, are to be preferred to scintillation crystals.

The electronics associated with any of the mentioned detectors generally consist of power supply, preamplifier, linear amplifier, and a scaler or a multichannel pulse height analyzer. All the electronic components must exhibit a high stability in order to exploit completely the resolution of the detector. Care should be taken to prevent temperature changes which can cause very noticeable drifting; an air conditioned room is therefore advisable. Nowadays there is no problem in finding electronic parts having the required characteristics of high stability and minimum noise. An extensive list of manufacturers of nuclear instrumentation in the U.S.A. is reported annually in *Science* (e.g., Vol. 158 A, no.3,804 A, November 28, 1967, pp.74–84).

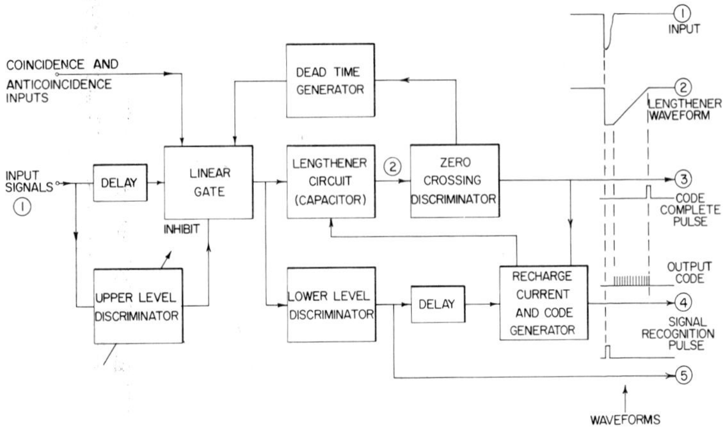

Fig.38. Block diagram of a typical encoder unit of a multichannel pulse height analyzer. (After GOULDING, 1965; courtesy of North-Holland Publ. Co.)

Pulse height analyzers are the most complex component and have undergone a substantial improvement in the last decade. Modern multichannel analyzers are wholly transistorized, with a magnetic core memory, typewriter, and/or punched tape readout and oscilloscope display. Although several different types are now available, most of them operate on the same principles. GOULDING (1965) has recently described the main types. A typical magnetic core multichannel analyzer consists in broad outline of two parts: the encoder (or converter) and the storage unit (or memory). Fig.38 shows a block diagram of the encoder, whose main purpose is to code the input signal into a digital form. This is accomplished by two main steps. The incoming signal is lengthened, in most cases by letting it charge a capacitor, which is then linearly discharged at a fixed rate. The time at which the capacitor voltage crosses the original zero level is taken as a measure of the amplitude (amplitude–time conversion). The capacitor discharge is then digitized by letting it gate a constant frequency oscillator, which gives rise to a number of pulses proportional to the duration of the discharge. In terms of greater details, the succession of events in the encoder is as follows (see Fig.38). An incoming signal, if it is not so high as to trigger the upper discriminator, is admitted through a linear gate to the lengthener circuit. Simultaneously it is examined by the lower level discriminator; if it is high enough to trigger it, a recharge current is sent with a short delay to the lengthener circuit

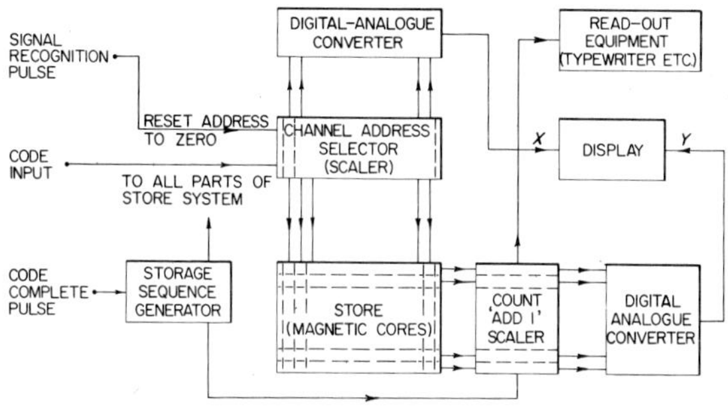

Fig.39. A block diagram of a magnetic core "memory" of a multichannel pulse height analyzer. (After GOULDING, 1965; courtesy of North-Holland Publ. Co.)

and a signal recognition pulse is generated (waveform *5*). During the recharge of the capacitor, the constant frequency oscillator is gated and the pulse train of the code output is generated (waveform *4*). After the capacitor has recharged to its original level, the zero crossing discriminator recognizes it and sends a code complete pulse (waveform *3*). During the processing of the signal no other pulse must be admitted; the dead time generator closes, therefore, the linear gate for all the processing time. Three pulses (signal recognition, code input, and code complete) reach the input of the storage unit (Fig.39). The heart of this unit is the memory or storage itself, usually consisting of a number of tiny ferrite toroids arranged in several planes, one above the other. The number of toroids in each plane represents the number of channels, while the number of planes determines the counts that can be accommodated in each channel, expressed in binary digits. Therefore, a 512 channel analyzer's memory which can store 2^{16} or 65,535 counts per channel, consists of 16 planes of 512 toroids each. The sequence of events in the storage unit is the following: the signal recognition pulse resets the address selector to zero, the code input steps it in correspondence of the column of toroids relative to the proper channel, and the complete code pulse adds the new signal to the channel through the "add + 1" scaler. The readout is made, after the counting has been completed, by stepping the address selector channel by channel, transferring the content of each channel to the "add + 1" scaler where it is used to operate a digital readout device (typewriter, punched tape, etc.).

Most of the multichannel analyzers also incorporate coincidence and anticoincidence circuits and allow the storage of separate runs in different sets of channels and, by transferring the contents of a group to another group, to perform addition or subtraction of spectra, e.g., background spectra.

Multidimensional pulse height analyzers have been recently developed allowing one to correlate events occurring in different detectors. In such analyzers channels are grouped in a plane bounded by two rectangular axes, *X* and *Y*. An example may clarify the way such analyzers work. Assume two detectors viewing the opposite sides of a ^{60}Co source and both connected to a 4,096 channel analyzer, having channels grouped in a 64 × 64 channels grid. Assume that the channel width is 50 keV and that no coincident pulses from detector *1* are recorded on the axis *X* and no coincident pulses from the detector *2* on the axis *Y*. It is known that ^{60}Co emits two gamma-rays (1.17 and 1.33 MeV) in cascade. When both gammas are totally absorbed in detector *1* (or *2*), two pulses will be recorded in channels *24* (1.17/0.05) and *27* (1.33/0.05) on axis *X* (or *Y*). But, if a 1.17 MeV gamma is totally absorbed in detector *1* and a coincident 1.33 MeV gamma in detector *2* (or vice versa), the two events will be recorded in one channel characterized by an *X* coordinate of *24* and a *Y* coordinate of *27* (or vice versa). Coincident events occurring in the two detectors can be easily resolved in this way from a complex spectrum. Fig.40 compares the spectra of a ^{60}Co–^{46}Sc* mixture; the unidimensional spectrum does not resolve the 1.118 MeV peak of ^{46}Sc from the 1.17 MeV peak of ^{60}Co, whereas they are clearly resolved in the multidimensional spectrum. As pointed out by NIELSEN and PERKINS (1967), this technique has a variety of applications in earth sciences, including low level determinations of spallogenetic radionuclides in meteorites (e.g., ^{26}Al, ^{46}Sc, ^{60}Co, ^{44}Ti, etc.) and radionuclides produced by spallation of atmospheric argon (e.g., ^{34m}Cl, ^{38}Cl, ^{38}S, ^{24}Na).

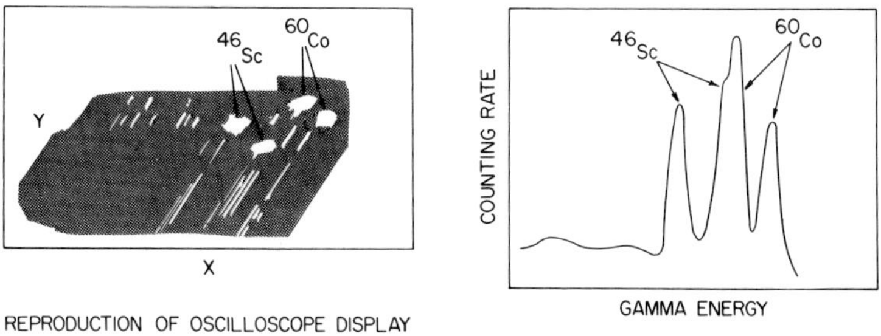

Fig.40. Comparison of multidimensional and unidimensional spectra of a ^{46}Sc–^{60}Co mixture. (After NIELSEN and PERKINS, 1967.)

The practical ultimate detection limit of all the above instrumentation is set by the background counting rate. The main sources of background in addition to the inherent activity of the detector (Chapter 2) are: (*1*) environmental radioactivity; (*2*) cosmic radiation.

The walls, ceiling, and floor materials of the laboratory always contain some natural radioisotopes, mainly members of the ^{232}Th and ^{238}U series and ^{40}K, whose radiations can penetrate the layer of air and reach the detector. The presence of ^{232}Th and ^{238}U gives rise to two short lived gaseous members of the respective series, ^{220}Rn and ^{222}Rn, and their decay products in the air. The contributions from all such natural impurities to the background can be reduced by using carefully selected material for the construction of

* ^{46}Sc emits two gamma-rays (1.118 and 0.885 MeV) in cascade.

the laboratory, by shielding the detector with a suitable absorbing material, and by anticoincident annuli.

WOLLENBERG and SMITH (1964) have surveyed the radioactivity of different rock formations in the western U.S.A. with the purpose of finding the most suitable material for a low background laboratory. Serpentinized ultramafic rocks (gamma specific activity: 0.0197 counts/min g; U = 0.11 p.p.m.; Th = 0.07 p.p.m.; K = 0.001%) appear to fulfill more closely the low background requirements. HEIER and CARTER (1964) and ROGERS and ADAMS (1969a,b) report lower values from unaltered peridotites and dunites from the U.S.A. as well as from South America, South Africa, U.S.S.R., and China. Calcareous rocks with low clay content generally have low potassium and thorium contents; WOLLENBERG and SMITH (1964) report average specific activities of California carbonate rocks to be only slightly higher (0.023 counts/min g) than serpentinized rocks. The same authors report data for Portland cements from several plants in the U.S.A. and Canada. Gamma specific activities of finished cements have been found to be about 36% higher than the raw materials, indicating an increase of U and Th concentrations during the manufacturing process.

The choice of suitable material is, of course, dictated by the cost, by the expected counting rate of the samples, and by the characteristics of the laboratory itself. For instance, the possible lack of ventilation in underground air conditioned laboratories can have the consequence of a continuous variability of the Rn content of the air. Therefore, in such cases, it may be more convenient to use material with a higher potassium content, but depleted in U and Th, or Ra deficient materials, such as cements made from Holocene oyster shells (ADAMS, 1964).

The simplest way to reduce further the background due to the local radioactivity is by surrounding the detector with a high purity shield. An inspection of the absorption curves reported in Appendix 1 shows that high Z materials are of course best fitted for this purpose.

It must be pointed out that the absorption curves in Appendix 1 have been evaluated in "good geometry" conditions, i.e., for a collimated beam of photons passing through the absorber and aimed at a detector. In that case, a high percentage of Compton scattered photons is actually removed from the beam and do not reach the detector. When photons are not collimated the loss by scattering of photons initially aimed at the detector is partially compensated by the scattering towards the detector of photons initially directed in other directions. As a consequence, in poor geometry conditions there is a build up of secondary photons reaching the detector. The ratio of the number of photons reaching a detector under poor geometry conditions to that under good geometry conditions is usually referred to as "build up factor"[1]. The single or multiple Compton scattering produces a degradation of the energy of the scattered photons, so that the spectrum of energy reaching the detector is shifted towards lower energies. Build up factors are higher for low Z materials.

Lead ($Z = 82$) is a frequent choice for shielding, because it has the combined advantages of high absorption coefficients, reasonable cost, and ease of casting and shaping. However, as a consequence of the occurrence of U in most lead ores, a radioactive isotope of the ^{238}U series, ^{210}Pb (RaD), is present in variable concentrations in recently purified lead. The purification process separates ^{210}Pb from the parent radionuclides so that it is no more supported and decays with its own half life (22.0 years) by β^- emission (accompanied by a low intensity gamma-ray) forming a short decay series having both β^- (^{210}Bi) and α (^{210}Po) activity. As ^{210}Pb cannot be isotopically separated at low cost, the only expedients are to use lead more than 100 years old, or to avoid contamination of lead ores that are naturally uranium free. Specific activities of lead refined at different times have been measured by several authors, including De Vries and Barendsen (1953), Grummitt et al. (1956); Grinberg and Le Gallic (1961); Weller (1964); and Loosli et al. (1967), and, as expected, lower activities have been frequently found in "aged" lead (Table V).

Iron ($Z = 26$) and steel are often used as shielding materials. However, the increasing use of ^{60}Co in the steel industry in recent years suggests the choice of steel smelted before 1945. Contamina-

[1] In addition to the above defined number build up factor, energy and dose build up factors are frequently used.

tion by fall-out fission products, e.g., ^{103}Ru, must be considered.

Mercury ($Z = 80$) has good absorbing characteristics and can be easily purified. Triple distilled mercury is extremely pure, but the cost is the main obstacle for large scale use. A serious drawback is the difficulty of building adequately pure containers.

TABLE V

SOME SPECIFIC ACTIVITIES OBSERVED IN LEADS OF VARIOUS AGES

Lead	*dis./min/g*	*Reference*
Modern	3.5	GRUMMITT et al. (1956)
Ordinary modern	90	GRINBERG, LE GALLIC (1961)
Specially refined modern	1	GRINBERG, LE GALLIC (1961)
18th century (Belgium)	56	GRINBERG, LE GALLIC (1961)
18th century	8	GRINBERG, LE GALLIC (1961)
17th century (Versailles)	4	GRINBERG, LE GALLIC (1961)
16th century	0.35	GRUMMITT et al. (1956)
Roman ships	4	GRINBERG, LE GALLIC (1961)

Charged and uncharged particles and photons hitting the shield cause the emission of secondary photons, mainly of low energy, which, in the case of small distances between shield and detector, can easily reach the detector. The interference of these secondary photons in the region of the spectrum under study must be carefully considered in the choice of the proper shielding material and shield geometry. Photoelectric absorption by the shield of photons coming from the sample or from all the sources of background causes the emission of K X-rays characteristic of the elements in the shield. A high fraction of these are photoelectrically absorbed by the detector, giving rise to sharp full energy peaks. K X-rays from low and intermediate Z materials interfere of course with the X-ray energy range, but K X-rays from high Z materials, such as Pb ($\sim$72 keV) or Hg ($\sim$70 keV) interfere with the low energy gamma spectrum. Moreover, charged particles passing through the shield give rise to the emission of bremsstrahlung X-rays. The energy fraction of incident monoenergetic electrons (the most significant source of bremsstrahlung) converted into bremsstrahlung increases with increasing atomic number of the absorber and increasing maximum energy of the electron. Bremsstrahlung radiations interfere mainly in the energy region below 0.5 MeV.

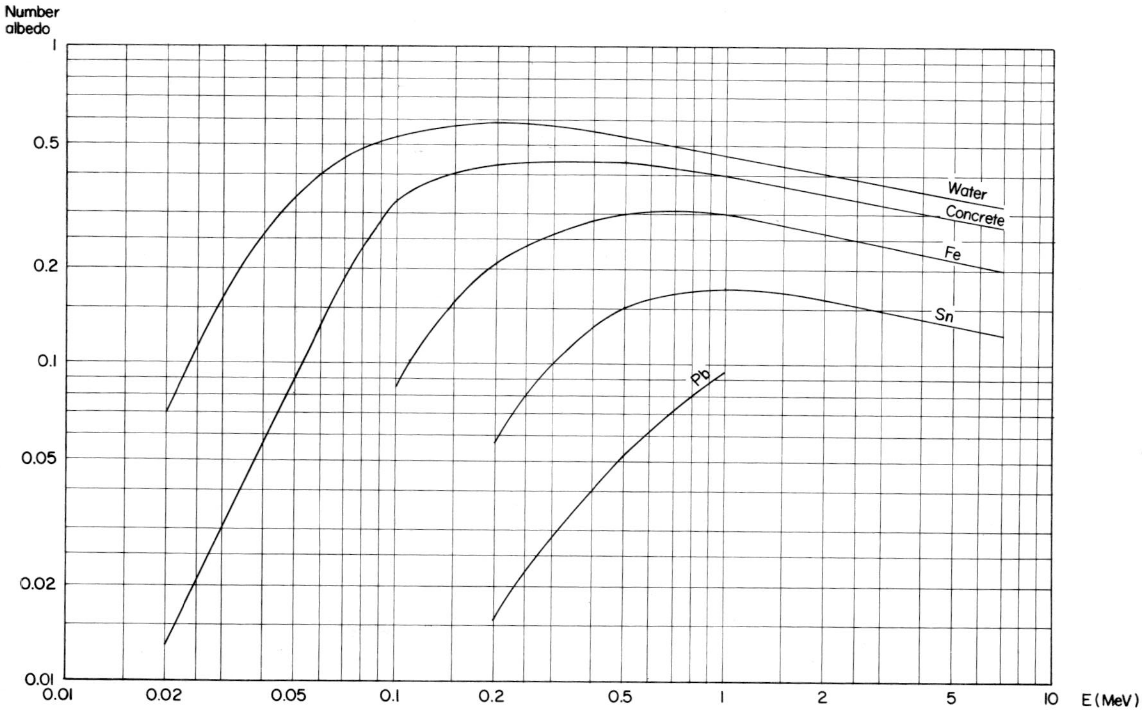

Fig. 41. Number albedo against photon energy for water, concrete, Fe ($Z = 26$), Sn ($Z = 50$), and Pb ($Z = 82$) absorbers. (After the data of BERGER and RASO, 1960.)

An additional interference arises from the backscattering by the shield of photons coming from the sample, and from all the radioactive impurities in the sample container and in the detector assembly. In Fig.41 is given the number albedo (i.e., the ratio of backscattered to incident photons) for materials of different Z, after BERGER and RASO (1960). The albedo increases at first with decreasing energy of the incident photons; it then reaches a maximum, and finally quickly decreases at very low energies. The increase is caused by lower energy photons because: (*1*) the penetration is more shallow and therefore the scattered photons have a higher probability of reaching the detector; (*2*) the fractional energy loss for collision is smaller; and (*3*) the differential cross section for backscattering is higher (Fig.10). On the other hand, photoelectric cross section increases rapidly with decreasing energy, so that a quick decrease occurs where photoelectric absorption becomes significant. The albedo is higher for low Z materials mainly because of their higher transparency (lower electron density) to the backscattered radiation. Fig.9 reports the energy of 180° backscattered versus that of incident photons; all the 180° backscattered radiation is below about 250 keV. Backscattered photons have, therefore, a high probability to be fully absorbed in the detector, and give rise to a broad peak in the spectrum. Fig.42 from HEATH (1957) shows the variation of the intensity of the backscattered radiation for Fe and Pb shields. An effective reduction of the backscattered peak is obtained by enlarging the dimensions of the space around the detector. A nearly complete removal of the radiation lower than 150–200 keV can be obtained by an air gap of approximately 50 cm between the shield and the detector. A further way to reduce the intensity of secondary photons from high Z shields is to interpose a thin low Z graded shield, e.g., 1 mm Fe and 1 mm Cu, or Cd, Cu, and Perspex, as suggested by WATT and RAMSDEN (1964). The material for the sample containers must be carefully checked for any undesirable activity, due to radioactive impurities or to secondary photons. Polyethylene or equivalent plastic containers are often adequate; aluminum containers can also be used provided that they have been carefully checked for ^{226}Ra impurities.

A further main source of background is the cosmic radiation.

Primary cosmic radiation consists of about 85% protons and 15% alphas plus a minor component of nuclei heavier than He. By interaction with the atmosphere they give rise to three main secondary components: high energy nucleons, charged pions and neutral Π^0 particles.

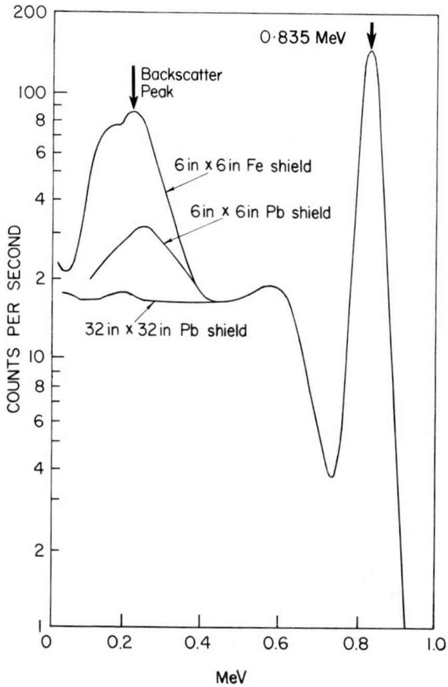

Fig.42. Relative increase of the backscattering peak for Fe and Pb shieldings and for 6 × 6 inch (15.2 × 15.2 cm) and 12 × 12 inch (30.5 × 30.5 cm) detector's rooms. (After HEATH, 1957.)

High energy nucleons, accounting for about 20% of the energy of primary radiation, interact in turn with the atmosphere by pion production, spallation, etc., starting the nucleonic cascade. It propagates down to sea level (where nucleons plus muons are about 20% of all particles) until capture in the resonance or thermal energy region occurs. Photons are produced either in the capture process or in inelastic scattering; the energy spectra of photons by capture of thermal neutrons by iron, concrete, lead, and soil are shown in Fig.43 (from MAY and MARINELLI, 1964). In lead, the capture of thermal neutrons gives rise to photons with energies sharply peaked at around 7 MeV; whereas thermal neutron capture by less dense and lower Z materials produce a lower energy spectrum.

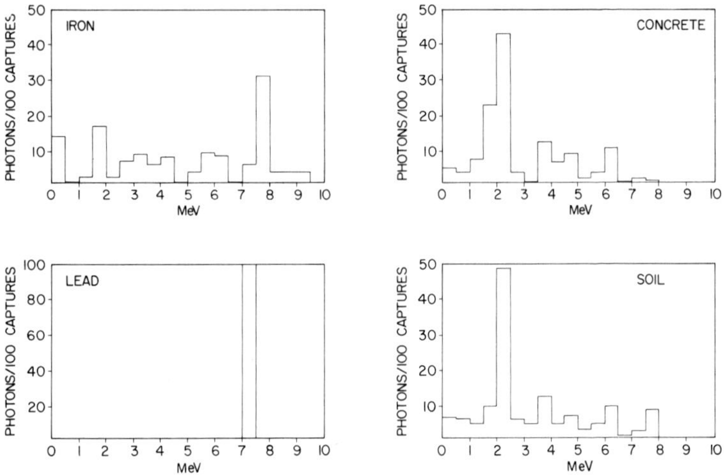

Fig.43. Energy spectra of photons emitted after capture of thermal neutrons in Fe, concrete, Pb and soil. (After MAY and MARINELLI, 1964; courtesy of Rice University.)

Charged pions, accounting for about 30% of the energy of primary radiation, decay shortly (mean life: 0.25 nanosec) to highly penetrating muons, constituting the "hard component" of the cosmic radiation at sea level. Muons, because of their high penetrating power, can reach sea level; they can disappear either by decay:

$$\mu^{\pm} \rightarrow \beta^{\pm} + 2\,{}^{0}n \quad ({}^{0}n = \text{neutrino})$$

or negative muons can be absorbed by nuclear capture (the Coulombic barrier of the nucleus forbids capture of positive muons):

$$\mu^{-} + {}^{A}_{Z}X \rightarrow {}_{Z-1}{}^{A}X \rightarrow n + {}^{A-1}_{Z-1}X + {}^{0}n$$

Capture probability increases with increasing atomic number of the absorber, being equal to decay probability for $Z = 11$. Positrons resulting from the process of decay give rise by annihilation and cascade to many intermediate energy photons, whereas neutrons from the capture process can produce photons after capture or inelastic scattering, as described above.

Neutral pions also decay quickly (mean life = 0.19 femtosec) giving rise to two high energy photons starting, by interaction with the atmosphere, the electron–photon cascade, which forms the "soft component" of the cosmic radiation at sea level. This component is easier to deal with; about 10 cm of Pb are sufficient to absorb electrons, although photons of a few MeV of energy can penetrate because of the broad minimum in the Pb absorption curve between the photoelectric and the pair production regions (see Appendix 1).

It is evident that the use of a high Z absorber around the detector is not always a means to reduce the cosmic ray background. On the contrary, showers of charged particles, neutrons, and photons are produced by the interaction of the most penetrating components (nucleons and muons) with the shield.

It has been observed that the shower particles, with the possible exception of neutrons, are all coincident within 1 μsec. Therefore, if a ring of counters is placed around the detector so that a good probability exists that they can detect at least one particle of the shower, any other particle of the shower absorbed by the main detector can be eliminated by the use of an anticoincidence circuit. Gas filled detectors, such as GM or proportional counters, are often used as anticoincidence counters but they have the serious drawback of a low detecting efficiency for photons. A higher photon detecting efficiency can be reached by the use of a large NaI(Tl) anticoincidence annuli, but their use is at present restricted to relatively small sizes of the main detector. Plastic scintillators are now available in very large sizes and can be successfully used as anticoincidence rings, having a satisfactory absorption cross section for intermediate energy photons and a high hydrogen content so that they also thermalize neutrons. However, it must be pointed out that the use of anticoincidence rings has the shortcoming of increasing the dead time of the entire system. A different approach can be the incorporation in the shield of a hydrogenous material, such as paraffin wax, to thermalize neutrons, impregnated with a material, such as boron, having a high neutron capture cross section and which does not emit appreciable capture gamma-rays.

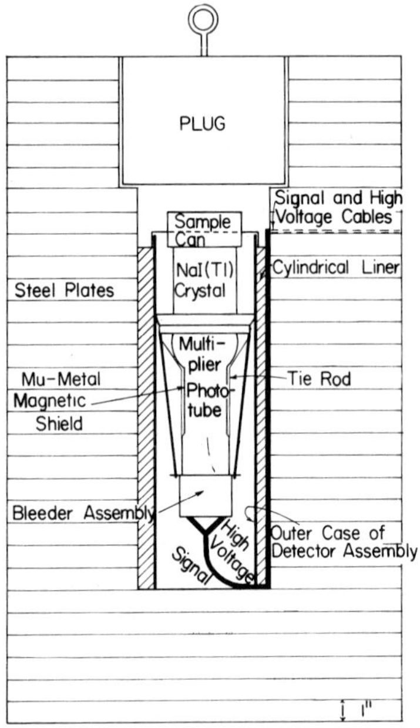

Fig.44. Detecting and shielding assembly for gamma spectrometry of common rocks, operating at the Department of Geology of Rice University, Houston, Texas. (After ADAMS, 1964; courtesy of Rice University.)

The cosmic ray contribution can be drastically reduced where the detecting assembly is placed in a deep cave. For instance, a decrease of 98.8% of the muonic background has been observed passing the detector (a proportional counter) from a sea level shielding of 320 g/cm^2 of concrete plus lead to a cave under 60 m of calcareous rock, corresponding to a shielding of 15,000 g/cm^2 (DELIBRIAS and REPAIRE, 1967).

The cost of adequate cosmic ray shielding is high and its use is therefore restricted to special cases, where less than a few counts per minute are expected. For geochemical work with ordinary igneous and sedimentary rocks it is often possible to get higher counting rates by using larger samples rather than by reducing drastically the cosmic ray background.

Typical arrangements for ordinary geochemical work are those operating at the Department of Geology of Rice University, Houston, Texas and at the Istituto di Fisica Terrestre of the Università di Napoli, Italy.

The gamma-ray spectrometry laboratory of the Keith–Weiss Geological Laboratories at Rice University (ADAMS, 1964) has been built below ground level, the surrounding soil having a nominal content of about 7–9 p.p.m. Th, 2.5–3.6 p.p.m. U and 0.8–1.2% K. The room has a concrete baffle entrance and its walls, ceiling, and floor are about 20 cm thick Ra deficient concrete. The total vertical shielding from the overlying three storied building amounts to about 65 cm of concrete. Two detecting assemblies are in use. A 3 × 3 inch NaI(Tl) Harshaw integral line is inside a steel shield composed of 25 mm thick and 450 cm^2 steel plates (Fig.44). The second one consists of two 3 × 3 inch NaI(Tl) Harshaw integral lines facing each other and housed inside a triple distilled mercury shield. No major reduction in the background level has been observed with the mercury shield, mainly because of the inherent activity of the welding material of the container for mercury. Welding flux rods usually contain some titanium dioxide which, if it has been obtained from placer sands, can easily be contaminated with small amounts of radioactive minerals.

A vacuum tube high voltage power supply with internal stabilization (Tracerlab RLI–7) supplies 900–1,100 V for the iron shield detector. The mercury shielded detectors have separate high voltage battery supplies. Two solid state Nuclear Data multichannel analyzers (a 256 channel, ND–102, and a 512 channel, ND–120) are connected to the detectors and the readout is on a typewriter or punched tape.

The sample can shown in Fig.44 is a commercially available 8 oz. can. An array for the gamma spectrometry of rock cores has been set up in connection with the mercury shield assembly. The core pieces are enclosed in plastic tubing and inserted and taken off from the detector by a custom-built automatic sample changer. A programmer for sample changing, data read out, and memory erasure allows one to analyze a large number of cores automatically. Powdered samples in plastic core cylinders ($1\frac{1}{4}$ inch or 3.2 cm inner diameter and 7 inch or 17.8 cm long) are also counted automatically. The limit of detection is 0.02 p.p.m. U or Th and 0.005% K (ADAMS, 1963).

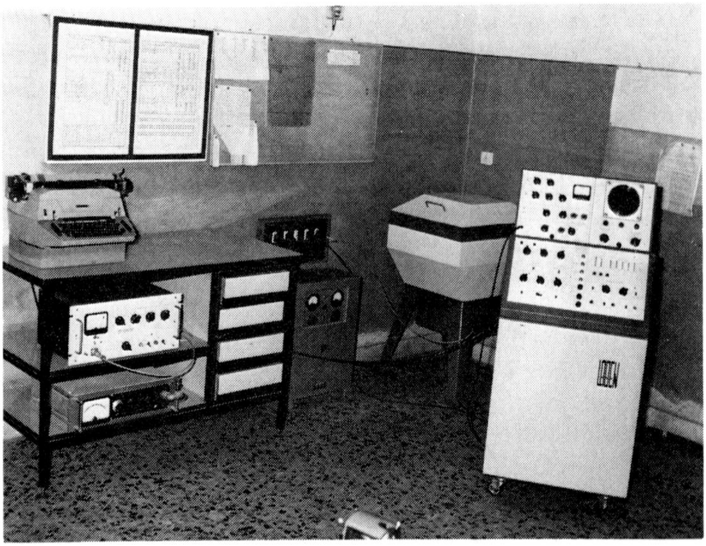

Fig.45. The gamma-ray spectrometry equipment for analysis of common rocks operating at the Istituto di Fiscia Terrestre of the Università di Napoli, Italy. From right to left: the 512 channel solid state magnetic core analyzer, the Pb shield enclosing the detecting assembly, the current stabilizer and, on the racks, the typewriter for the readout, a low voltage (for the cathode follower) and a high voltage (for the phototube) high stability supply. (After IMBO' et al., 1965.)

The gamma-ray spectrometry laboratory of the Istituto di Fisica Terrestre of the Università di Napoli is located on the third floor of a building, at about 15 m above ground level and 45 m above sea level. In this case, the high activity of the basement rock, a volcanic tuff with a content of about 30 p.p.m. Th, 20 p.p.m. U and 7% K mitigates against an underground installation. A high sensitivity gamma detecting assembly is in use (BONASIA and GASPARINI, 1965) consisting of a 3 × 3 inch NaI(Tl) Quartz and

Silice manufactured integral line assembly inside an "aged" lead shield, manufactured by SELO, Milan. The shield is composed of 5 × 10 × 20 cm and 5 × 10 × 15 cm lead bricks and has a minimum of 10 cm and a maximum (from zenith to 60°) of 15 cm overall thickness. The space for the detector has a volume of 4.5 l and the inner shield walls are lined with 1 mm Fe and 1 mm Cu in order to reduce the intensity of low energy photons from the shield. The total weight is 1,140 kg. The phototube is supplied 900–1,200 V by a high stability vacuum tube power supply (Italelettronica) and the pulse height analysis is performed by a solid state magnetic core 512 channel Laben analyzer with oscilloscope display and type-writer readout. Fig. 45 shows an ensemble view of the equipment (from IMBO' et al., 1965).

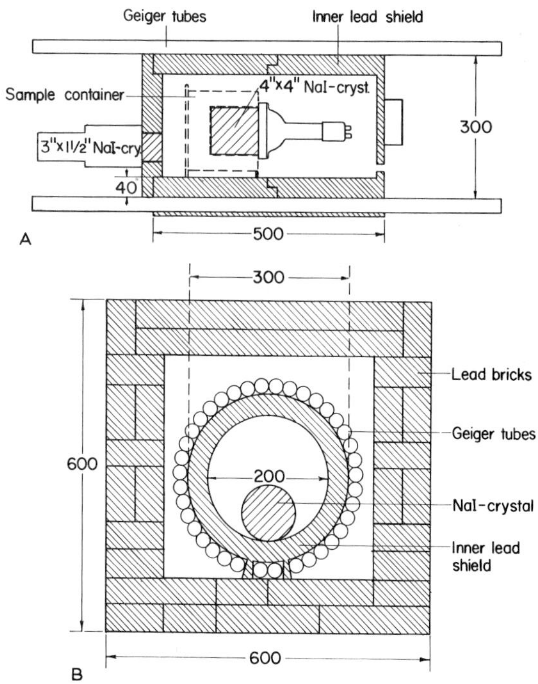

Fig.46. Cross section (B) and longitudinal view (A) of the low background equipment operating at the Physikalisches Institut of Heidelberg, Germany. The second NaI(Tl) crystal reported in the longitudinal section is used for γ–γ coincidence measurements. All the dimensions are in mm. (After SONNTAG, 1967.)

Thin aluminum sample containers are used, which surround all the crystal but the optical coupling side (the so-called "Marinelli hat"); these containers hold about 1 kg of crushed rock in a uniform layer about 2 cm thick. A comparable background is obtained in both laboratories, with typical counting rates of about 10 counts/min in the ^{40}K channels and about 15 counts/min in the energy range from the ^{40}K peak to 2,615 keV. More sophisticated low background laboratories using anticoincidence shielding annuli have been described recently by SONNTAG (1967) and NIELSEN and PERKINS (1967a).

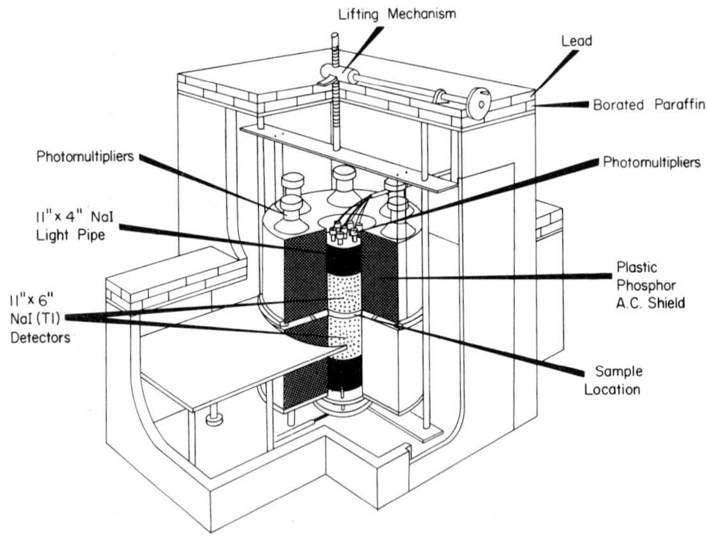

Fig.47. The low background equipment operating at the Battelle Memorial Institute, Richland, Washington. (After NIELSEN and PERKINS, 1967.)

The equipment described by Sonntag (Fig.46) and operating at the Physikalisches Institut of the University of Heidelberg, Germany, consists of a Harshaw 4 × 4 inch (10.2 × 10.2 cm) NaI(Tl) crystal inside a 4 cm thick selected lead inner shield, a GM anticoincidence ring and an outer 10 cm thick ordinary lead shield with moderate activity. The purpose of the inner lead shield is to reduce the number of photons coming from: (*1*) radioactive impurities in both the GM counters and the outer lead shield; or (*2*) radioactivity produced by muons in the outer shield. The total background in the whole energy range above 30 keV is 120 counts/min.

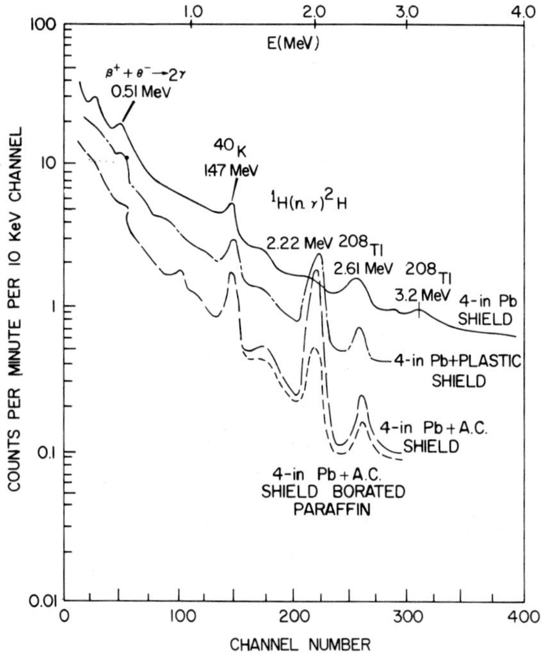

Fig.48. Background spectra obtained with the equipment of Fig.47 (lowest curve) and with other simpler arrangements. (After NIELSEN and PERKINS, 1967.)

The detector system operating at the Battalle Memorial Institute, Richland, Washington and described by NIELSEN and PERKINS (1967) (Fig.47) consists of two Harshaw 11 inch (27.9 cm) diameter by 6 inch (15.2 cm) high NaI(Tl) crystals, with 4 inch (10.2 cm) long pure NaI pipe facing each other. Each crystal is surrounded by a 36 inch (91.4 cm) diameter by 18 inch (45.7 cm) high plastic phosphor anticoincidence shield; each phosphor is viewed by six 5 inch (12.7 cm) diameter RCA 8055 multiplier phototubes. The whole assembly is housed inside a shield of lead and borated paraffin. The overall shielding is 10 cm lead in all directions and 10 cm of paraffin on the sides and on the top. The background reduction obtained with different variants of this system is shown in Fig.48. The background spectrum obtained by using a 10 cm lead shield and letting the plastic phosphors act only as a bulk shield is two or three times reduced in respect to that obtained by using a mere 10 cm lead shield. When the plastic phosphor ring is operated in anticoincidence the background is further reduced. The presence of the plastic phosphor ring gives rise to a prominent 2.225 MeV peak resulting from ^{1}H (n, γ) ^{2}H reaction. This peak has been reduced by lining the shield with about 10 cm of borated paraffin (5% wt. B).

The use of a plastic phosphor anticoincidence ring has the further advantage to reduce the height of the Compton spectrum, by cancelling all the pulses due to photons interacting with both NaI(Tl) and plastic detectors.

The detecting assembly is connected with a 4,096-channel multiparameter analyzer (Nuclear Data or Packard Instruments Co.) with photographic and paper tape readout.

The problems of attaining a low background have been discussed in more detail by many authors including PRICE et al. (1957); ARONSON and FRENCH (1962); BLIZARD (1962); FANO et al. (1962); WATT and RAMSDEN (1964); LEIPUNSKII et al. (1965).

FIELD SURVEYS

The gamma activity of the earth's rocks, waters, and atmosphere can be directly surveyed by measurement either at the very surface or by means of airborne or satelliteborne instruments (remote sensing). Furthermore, the gamma activity of rock formations inside the earth's crust can be tested by measurements in boreholes and wells.

The instrumentation needed for all such measurements must meet some common requirements. It must be very compact, rugged, light weight, with a low power consumption and a detector with high intrinsic efficiency and response.

Surface field surveys

Surface radiometric surveys can be carried out either by manborne or by vehicleborne instruments. Gamma activities are usually measured by GM or scintillation detectors.

GM detectors are very compact and require very simple electronics. A standard portable instrument consists of the GM tube connected by a short cable to a ratemeter with changeable scale ranges and integration times, and of a battery operated power supply. The integrating nature of ratemeters makes them useful mainly when a recording is desired. Each pulse coming from the detector is electronically converted into a charge ΔQ which is added to the charge Q of a capacitance C, shunted by a resistance R. The overall charge of the capacitor is proportional to the counting rate A and can be read by measuring the intensity of current (i) in the shunt:

$$i = Q/RC = \Delta Q A \tag{3.1}$$

A very compact portable ratemeter has been designed by GOLDSWORTHY (1960). It is wholly transistorized, weighs less than 2 kg, measures 10.2 × 10.2 × 22.9 cm and includes a battery power supply which can supply up to 1,500–1,800 V. Therefore, it can be used with GM as well as with scintillation counters.

The time needed for a reliable reading is strongly dependent on the counting rate, and consequently, on the intrinsic efficiency of the detector, on the geometry, and on the activity of the rock. When it is not possible to drill shallow holes in the surveyed rock, the detector is usually placed at a distance from the topographic surface to minimize the effects of small surface inhomogeneities (about 10–50 cm). In these conditions, it has been found that, using a 44.9 cm long by 28 cm diameter cylindric stainless steel GM tube, a reliable reading can be obtained after 8–10 m on a standard rock formation having about 14 p.p.m. U, 28 p.p.m. Th, and 4.5% K_2O (GASPARINI et al., 1961).

Portable scintillation detectors have much higher intrinsic efficiency and, hence, a reliable reading can be obtained in a shorter time. NaI(Tl) and CsI(Tl) crystals are also currently employed in most field work. A typical portable scintillation detector is described by WOLLENBERG and SMITH (1964). It consists of a 3 × 3 inch

(7.6 × 7.6 cm) NaI(Tl) crystal viewed by a 3 inch (7.6 cm) diameter DuMont phototube. The whole detection assembly is housed in a 12.7 cm diameter by 30 cm long thin walled stainless steel case and is connected by a 1.5 m long cable to an indicator unit of the type described by GOLDSWORTHY (1960), with four linear scale ranges up to 50,000 counts/sec.

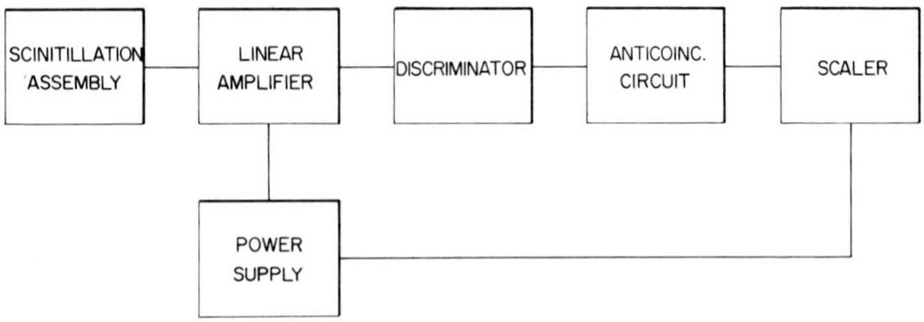

Fig.49. Block diagram of the Mark II portable gamma scintillation spectrometer. (After ADAMS and FRYER, 1964.)

Detectors can also be connected to portable scalers. Scalers count electronically single pulses coming from the detector. The circuit can be arranged in such a way that there is only one output pulse for each 2, 2^2, 2^3, ... 2^n (binary scale), or for each 10, 10^2, 10^3, ..., 10^n (decimal scalers). The number at the exponent represents the number of stages. The output pulse operates a mechanical register, and in some cases the interpolated pulses are registered by a system of glow lamps. A scaler–timer suitable for portable instruments has been designed by LÖVBÖRG (1967a). Digital ratemeters can also be used (G. E. Fryer and J. A. S. Adams, personal communication, 1969). A digital ratemeter is an automatically recycling scaler with a digital storage unit and a digital to analog converter. After one counting interval, the pulses accumulated in the scaler are transferred to the digital storage. The scaler is reset and begins a second counting interval with a negligible delay. The count stored in the storage unit operates the digital–analog converter and may be used to drive a strip chart recorder. A record will be obtained, where the counts accumulated in each counting period are displayed in forms of histograms with intervals equal to the counting period.

The development of scintillation detectors and transistorized circuits has allowed the construction of very compact and light field spectrometers, such as that described by ADAMS and FRYER (1964). Fig.49 is a block diagram of this instrument: it consists schematically of a 2 inch (5.1 cm) diameter by 3 inch (7.6 cm) high CsI(Tl) viewed by a 2 inch (5.1 cm) diameter photomultiplier, of a linear amplifier, a differential pulse height discriminator, an anti-coincidence circuit and a scaler; all the power is supplied by a 18 V Cd–Ni battery pack. Fig.50 shows the whole equipment ready for use. All the electronics are transistorized, the upper and lower discriminator levels are independently adjustable and the scaler readout is made by means of a 4 digits mechanical register. The pulse height analysis is made possible by the anticoincidence circuit allowing the passage of signals having sufficient amplitude to trigger the lower but not the upper discriminator level. The discriminator levels can be adjusted in such a way to explore successively the full energy peaks regions of ^{208}Tl (2.615 MeV), ^{214}Bi (1.76 MeV), and ^{40}K (1.46 MeV). The whole instrument pack weighs 36 kg, of which three-fourths are due to the portable Pb collimator. Other portable gamma spectrometers have been described by BARBEZIER et al. (1958); LÖVBÖRG (1967b); SCHNEIDER and SCHWERDTEL (1960), and DOIG (1968).

An absolute calibration of field instruments can be carried out only by reproducing as closely as possible the field conditions. A calibration platform can be built by scattering uniformly a known amount of U, Th, Ra, or K salts in a low radioactivity substance (e.g., concrete). The platform should be only some 10 cm thick and its size is dependent on the solid angle viewed by the detector. A field calibration could be carried out in a very flat area where the total activity is known with enough accuracy. The reproducibility of the geometry during the calibration and the field measurements can be assured by a portable lead collimator placed around the detector (ADAMS and FRYER, 1964). Some workers attempt to evaluate corrections for departures from 2π geometry. An accurate determination of the background due to instrumental impurities can be carried out by comparison with other detectors in a very low background environment, such as heavily shielded rooms, deep salt mines, etc. The instrumental plus cosmic ray background can be obtained by measurement on deep lakes or on the sea. In general terms the cosmic ray background is only the counting rate equivalent of a few tenths of p.p.m. Th or U as measured in the 2.62 MeV and 1.76 MeV photopeaks respectively.

A comparison of field and laboratory measurements has been carried out in northern New England soils by teams from the Department of Geology of Rice University and from U.S.A.E.C. Health and Safety Laboratory (ADAMS and LOWDER, 1964). The two teams obtained acceptable precision on thorium and potassium concentrations, although the field values were consistently somewhat lower. For uranium there was a deficit in the field measurements ranging from 10 to 50% of the U found in the laboratory. The explanation for these systematic differences between laboratory and field lies, partly, in the fact that the field measurements refer

Fig.50. Views of the Mark II portable gamma scintillation spectrometer: (A) in position to take to field stations; (B) disassembled to show the major sub-assemblies. (After ADAMS and FRYER, 1964; courtesy of Rice University.)

by definition to in situ soil material, whereas the laboratory determinations are usually in terms of weight unit of dry soil. A water content of 10–20% by weight in the soil is not unusual and would produce some of the observed effects. The larger observed differences in the uranium concentrations are also related to the degree of equilibrium of ^{222}Rn with its parents; the in situ situation is probably characterized by a more severe disequilibrium. Finally, it is doubtful in many cases that the laboratory sample is fully representative of the in situ system.

Fission products from nuclear explosions can be a serious cause of interference in field work (see Chapter 4 for the main interferences). Intermediate and long half life fission products are most responsible for soil activity for about a month after the explosion. They can be quite uniformly deposited by settling through the air or can be transported by rainwater and concentrated in low places where rainwaters are collected. A typical case occurred during a gamma radiometric survey of the bottom of the Solfatara volcanic crater, near Napoli, Italy, in the spring of 1962. Counting rates higher by a factor of 2.5 above normal were found in some depressions were rainwater collects. The counting rate went down to normal when the detector was placed in a shallow hole. Laboratory gamma spectrometry of a soil sample showed that such activity was due to ^{95}Zr, ^{95}Nb, and ^{144}Ce (see Appendix 2) probably from the 50 Megaton nuclear test in New Zemlia in the autumn of 1961 (GASPARINI et al., 1963). Accumulations of ^{144}Ce, ^{103}Ru, ^{95}Zr, ^{95}Nb, ^{134}Cs, and ^{140}La were observed in that same place (see Appendix 2) after the Chinese nuclear tests in 1965–1966 (GASPARINI and LUONGO, 1965). Similar effects were observed on the Rice University campus in 1965 (ADAMS, 1966). Fission products are usually confined to a very thin surface layer, or along cracks and fissures, and measurements in very shallow holes can be helpful, in dubious cases, to distinguish natural radioactivity from fall-out.

When field stations can be reached by vehicle the weight of the instrumentation is no longer a major problem and a multichannel analyzer can be taken into the field. A system of this type has been adopted by LOWDER et al. (1964) and BECK et al. (1964a,b). The equipment consists of an unshielded 5 inch (12.6 mm) diameter by 3 inch (7.6 mm) high NaI(Tl) detecting assembly mounted on a small wooden tripod facing downward through a hole in the wooden platform. A 1.50 m long cable connects the detector with a 256 channel pulse height analyzer which is mounted up in the vehicle. A converter supplies 116 V–60 cycle a.c. from the 12 V d.c. automobile battery.

Total gamma traverses can be carried out by vehicle-borne instruments. In this case a recorder is connected to the indicator unit, the drum being in some way connected to the wheels of the car, so that its rotating speed is proportional to the speed of the car. The detector can be placed either outside the car or it can view the ground surface through a hole in the bottom of the vehicle. Useful information from a geochemical point of view can be obtained only when traverses through fields are possible, otherwise the main contribution to the counting rate comes from the road materials.

Airborne surveys

Several research and development groups have published the

results of airborne gamma spectrometric surveys. DARNLEY (1968); DARNLEY and FLEET (1968); DARNLEY et al. (1968) have described in detail their surveys from both helicopters and fixed wing aircraft (see Fig.51). These workers found considerable promise in using gamma spectrometric surveys to delineate rock formations and the structural relationships among rock formations with different gamma spectrometric "signatures" (see Fig.51). FOOTE (1968a)

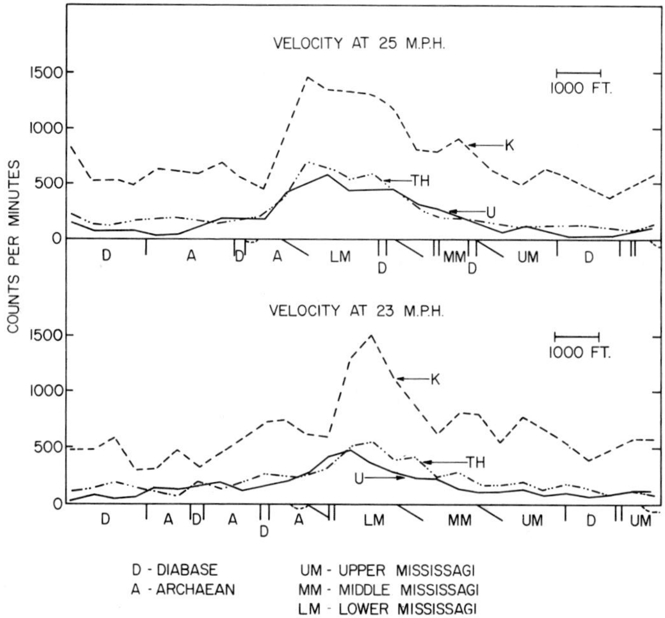

Fig.51. Helicopter borne radiometric profiles at 500 ft. (170 m) terrain clearance, Elliot Lake test strip. The detector assembly consisted for three 5 × 5 inch (12.7 × 12.7 cm) NaI(Tl) crystals. (After DARNLEY and FLEET, 1968.)

has described the delineation of soil types by airborne gamma spectrometry. FOOTE (1968b) has also described some of the complications that ^{222}Rn and its daughters can cause for airborne gamma spectrometry. The Aerial Radiological Monitoring Surveys (ARMS) have been reported by GUILLOU (1964) and HAND (1964); these surveys have covered hundreds of thousands of square miles in order to determine the broad general natural gamma-ray background around nuclear installations. VINOGRADOV et al. (1967) have described the Soviet Automatic Station Luna-10, which orbited a gamma-ray spectrometer around the moon in 1967. The gamma spectrometric data from Luna-10, although not of the highest quality, did reveal important information about the composition of the lunar surfaces and some of the processes taking place on the lunar surface (see Chapter 7, p.217). The Luna-10 experiment was a special case in that the moon has no atmosphere to attenuate gamma-rays in the manner that the earth's atmosphere does (see Fig. 52). ADAMS (1968) has described gamma spectro-

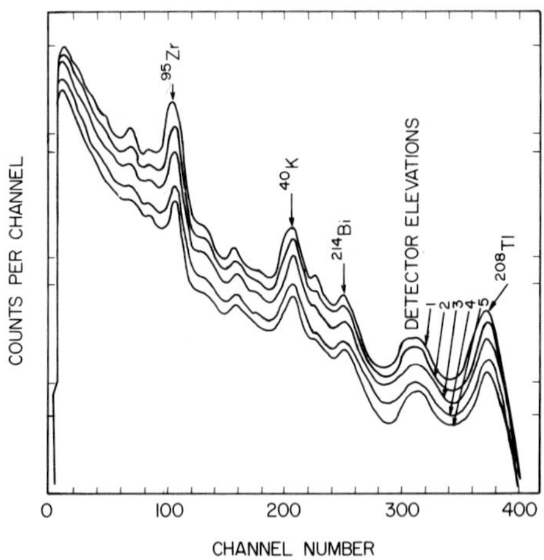

Fig.52. Gamma-ray spectrum of the earth as a function of the detector elevation above ground level; curve *1*:6 ft (18 m); *2*:50 ft (17 m); *3*:150 ft (51 m); *4*:350 ft (117 m); *5*:500 ft (170 m). The detector was an $11\frac{1}{2}$ inch (29.2 cm) by 4 inch (10.2 cm) NaI(Tl) crystal. (After PURVIS and FOOTE, 1964; courtesy of Rice University.)

metric surveys of beach sands and granites with a helicopter-borne spectrometer.

In evaluating the effectiveness of gamma spectrometric surveys one should note that the question is not just that of an instrument, but of a balanced design of spectrometric instrumentation, logistical support, ground speed, altitude above the ground, location on the ground, and, above all, the nature of the geological target being sought. It is not likely that a single airborne gamma spectrometric system will be ideal for every application. Indeed, it is quite possible that for many applications in mineral exploration two systems would be ideal; thus, a broad regional survey with a fixed wing aircraft carrying a large volume of scintillating crystal on predetermined flight lines at maximum altitude might well be followed by a helicopter system to delineate with greater resolution positive and negative anomalies using a smaller volume of crystal, live time readout and reduction of data, and the ability to land and collect samples, if such collecting is justified.

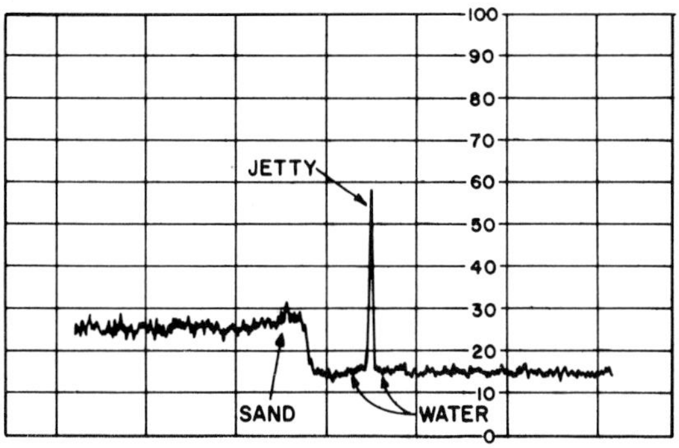

Fig.53. Survey of Galveston, Texas, jetty breakwater made at 50 ft (17 m), 60 knots, 500 counts full scale, 0.15 to 3.0 MeV and 0.5 sec time constant. The divisions along the abscissa are 1 inch (2.54 cm) of the original record, and the chart speed was 4 inch (10.2 cm) per minute. (After ADAMS, 1968.)

ADAMS (1968) has quantitatively evaluated the ground resolution of a helicopter system with a 5 inch (12.7 cm) by 5 inch (12.7 cm) diameter NaI(Tl) crystal flown at 50 ft. and 60 knots. Fig.53 shows the record obtained crossing a granite block jetty half a mile from shore with a 0.15–3.0 MeV window. ADAMS (1968) reports: "Twenty crossings of granite jetties indicate that the actual width of about 12 ft. gave an apparent anomaly width of about 190 ft. (= 0.125 inches on the record). Taking the jetty to be an infinitely thin line source, it is estimated that on flat terrain at 50 ft. and 60 knots the area primarily (>90%) under investigation at any one instant has a radius of approximately 75 ft. and that during a single 0.5 sec time interval, the center of this circle of investigation is advanced 50 ft. in the direction of traverse "

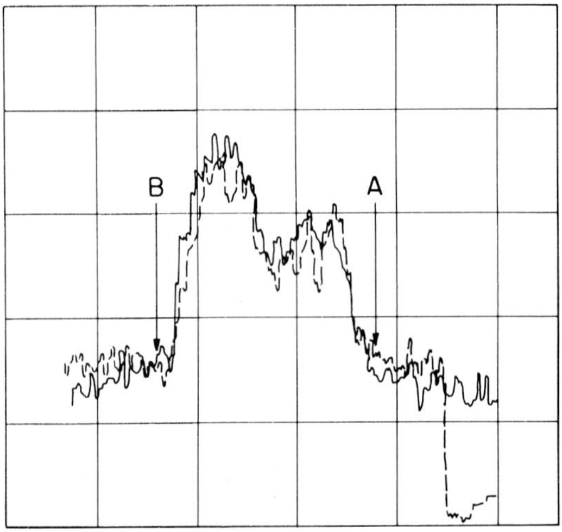

Fig.54. Surveys of Independence Rock on successive days, flying from *A* to *B* on the first day and from *B* to *A* on the second day. Experimental conditions are the same as those described for Fig.53. (After ADAMS, 1968.)

ADAMS (1968) also demonstrated that in flying at 50 ft. over rough terrain, it was possible to obtain quite satisfactory precision (see Fig.54). Quantitative estimates of thorium to ±3 p.p.m. can be obtained for granite and beach sands. Equally quantitative estimates for uranium and potassium have not, as yet, been reduced to practice. In calibrating the helicopter system, a major difficulty is the size and heterogeneity of the rock volume under investigation during each 0.5 sec time interval. It is useful to note the relative scale of measurements in the laboratory compared to the collimated surface instrument, compared to the helicopter, compared to the large fixed wing systems. Taking the laboratory scale as one, the relative sizes are approximately 1 : 3,000: ≈ 6,000,000: ≈ 120,000,000 respectively. It follows that the helicopter calibration in absolute units, although difficult, can be reduced to practice most easily by use of the thoroughly calibrated Mark II collimated instrument (see pp.90–93).

Airborne gamma spectrometric systems are expensive in terms of original equipment and operating costs, particularly in the case of helicopters and large fixed wing aircraft. On the other hand, the costs per data point or line mile flown is very low in comparison to alternative methods. It is important to note that the airborne surveys are independent of ground trafficability.

Borehole logging

Gamma borehole logging is at present a major technique of exploration and evaluation of uranium ores. DODD et al. (1967) have estimated that at least 6 million cubic ft. of rock have been surveyed by borehole logging in the western U.S.A. during 1967. Borehole logging furnishes also a rapid and direct means for the study of the subsurfacial distribution of radioactive elements, which study is most helpful for the interpretation of terrestrial heat flow data (see Chapter 7, pp.208–213). A further application is the stratigraphic correlation of subsurface rock formations.

Gamma borehole logging for uranium exploration is usually carried out in conjunction with auxiliary techniques which allow the determination of other physical parameters of the surveyed formations. Single point resistance, spontaneous potential, and gamma–gamma (density) are the most used of other logging techniques. Borehole logging methods used for exploration and evaluation of uranium ores have been recently reviewed by DODD et al. (1967).

Gamma logging is usually carried out in uncased holes having diameters in the range 5–25 cm (2–10 inch). The influence of deviations of the hole diameters from the nominal value on the detector's response is higher the smaller the diameter of the hole. This influence becomes particularly noticeable for holes smaller than about 15 cm (6 inch) diameter. Logging holes are ordinarily drilled to depths in the range 70–200 m (200–600 ft). Depths deeper than 800 m (2,500 ft.) are rare, but becoming more common. As a consequence, detectors are usually not exposed to severe temperatures and pressures. During logging, holes are usually either filled with air or water and mud.

The particular environments where detectors operate and their great distance from the surface electronics give rise to some particular problems. In fact, the need of operating the gamma-ray detector at the end of a long cable requires that the high voltage supply is best included with the detecting assembly. Furthermore, it is best also to locate the preamplifier close to the detecting assembly. Because of the small size of the logging holes, such units must be remarkably compact in the horizontal dimensions. Additional problems arise if a spectrometric determination is required. In this case, in fact, the proportionality between the photons' energy and pulse height must be maintained through tens and sometimes a hundred meters of cable. Furthermore, quite pronounced drifts are to be expected as a consequence of rapid temperature changes within the hole; gain stabilization is therefore required for best results.

A typical instrumentation for gamma borehole logging is that used by the Grand Junction Office of the U.S. Atomic Energy Commission, which has been described by DODD (1966) and DODD et al. (1967). All such units are truck mounted and primary a.c. power supply is given by a 4 kW generator. Two probes are in common use: one having a $\frac{1}{2} \times \frac{1}{2}$ inch (1.3 × 1.3 cm) NaI(Tl) and the other one having a 5 × 4 inch (12.7 × 10.2 cm) NaI(Tl) crystal as a detector. The smaller crystal is used to measure ore-grade activities, in order to prevent excessive overloading of the pulse transmission. The 5 × 4 inch crystal is used for normal rock activities or when spectrometric determinations are performed, in order to obtain an adequate counting statistics at a reasonable logging speed. Each crystal–multiplier phototube assembly is connected to a logging cable through an amplifier and a cable driver. The logging cable is a double-armoured three conductor cable and its surface end is connected to a hybrid cathode-follower, which drives a linear amplifier. Finally, signals enter either a single channel pulse height analyzer or a digital ratemeter. Analog and digital readout is simultaneously provided. Interval control signals for the digital readout are developed in the synchro-driven depth measuring system through a light activated pulse generator. A 1,024 channel pulse height analyzer is also available for gamma-ray spectrometry. In this last case, digital and analog readout are both possible.

Auxiliary instrumentations for electrical and density logging are also commonly mounted on the same truck.

An important problem is the absolute calibration of the logging instrumentation. SCOTT et al. (1961) have developed a method allowing one to correlate the gross counting rate obtained in U ores to the percent equivalent U_3O_8 by weight (radiometric grade). They have demonstrated both theoretically and experimentally by full scale hole models that, if the logging speed is constant, the following equation applies:

$$\bar{g}_\gamma x = \mathrm{k}A \tag{3.2}$$

where $\bar{g}_\gamma$ is the radiometric grade of the uranium ore, x is the thickness of the ore, A is the corrected positive anomaly area in the logging curve, and k is a calibration constant. The constant k must be determined empirically by full scale models for each hole, for each probe and for each logging and analytical procedure. The proportionality expressed by eq.(3.1) has been further confirmed by calibration with ores of known ore grades and thicknesses. The average radiometric grade is evaluated by dividing $\bar{g}_\gamma x$ by the ore thickness x. The ore thickness is the distance between the ore boundaries, which can be considered to be proportional to the distance between the two half amplitude points on each side of the anomaly. Corrections must be applied for system nonlinearity and for variations of the logging environment from the standard calibration conditions. The radiometric grade, $\bar{g}_\gamma$, is most useful and correct for U evaluation in situations where there are a large number of data points on a deposit of economic dimensions. In other words, one may not have secular radioactive equilibrium in a small sub-sample of a deposit that overall is near secular equilibrium.

Spectrometric determinations in down-hole conditions are complicated by the special nature of the gamma spectrum obtained from the practically infinite size of the volume under consideration. The main complication is that the number of photons Compton scattered before reaching the detector is greatly increased. The shape of the down-hole spectrum is deformed in that the low energy/high energy ratio is much higher than that from a small sample obtained under laboratory conditions. A much more pronounced Compton continuum masks nearly completely any full energy peak occurring at energies lower than 1.0 MeV. An additional characteristic of the down-hole spectrum at low energies (< 0.5 MeV) is due to the increased probability of photoelectric absorption of photons by high atomic number elements, including U, in the rock. The photoelectric absorption of monoenergetic photons from the rock has been proposed as a method for evaluating the composition and average atomic number of the rock (see CZUBEK, 1965). Methods of measuring the equilibrium conditions of the members of the ^{238}U series with the ratio of the low energy and the high energy parts of the spectrum (see Chapter 4 and Table XXIV) have been proposed (see DODD et al., 1967).

Gamma–gamma logging is an auxiliary technique which permits one to estimate the density of subsurface rock formations. This method has been first described by FAUL and TITTLE (1951). The principle is quite simple. If a source of monoenergetic gamma-rays is used to irradiate a given rock formation, and if the photon energy is chosen so that the Compton scattering is the dominant absorption process, the fraction of photons which will be Compton scattered by a given angle is controlled by the electron density of the scattering medium (in our case, the rock). If n_e is the electron density of the scattering medium and ρ its bulk density, we have:

$$n_e = N(Z/A)\rho \qquad (3.3)$$

where N is the Avogadro number and Z and A are the atomic number and mass number of the scattering medium. As the Z/A ratio is approximately constant for most rock forming elements (about 0.5, neglecting hydrogen) the intensity of Compton radiation scattered by a given angle is ultimately proportional to the bulk density of the scattering medium (for more details see PICKELL and HEACOCK, 1960; TITTMAN and WAHL, 1965; DODD et al., 1967; HOMILIUS and LORCH, 1967).

DODD and DROULLARD (1965) have developed a method which overcomes the problem of overshadowing of scattered photons by an intense natural gamma field (e.g., like that from uranium ore). This method also allows one to evaluate the wet bulk density of a rock. It is essentially based upon the difference of intensity between two beams of photons scattered at different angles, e.g., at different source-detector distances within the probe. Probes for gamma–gamma logging contain, one above the other, a suitable gamma-ray source (e.g., ^{60}Co) attached to a collimator, a lead shield to protect the detector against direct irradiation from the source, and a detector, which in most cases is a NaI(Tl) crystal. The detector–source distance can be changed by a motor, controlled from the surface, which drives a device for positioning the source–collimator assembly. Calibrations are here again carried out with full scale models; the density of the material surrounding the hole model is determined by other physical methods and corrected to a Z/A ratio of 0.5.

CHAPTER 4

Gamma Ray Emitters

NATURAL EMITTERS

Natural unstable or radioactive isotopes can be classified as "primevals" and "secondaries."

Primeval radioisotopes are those having sufficiently long half-lives to have survived in detectable amounts since the time of nucleosynthesis. The shortest lived known isotope of this category is ^{235}U with a half-life of 0.71 gigayears. Many primeval gamma emitters are known and Table VI lists most of them, together with their half lives, natural isotopic ratios, average crustal abundances, and gamma activity in average crustal material. ^{40}K is by far the biggest contributor, accounting for about 98% of the total gamma

TABLE VI

MAIN PRIMEVAL NATURAL GAMMA-RAY EMITTERS

Isotope	*Isotopic ratio (%)*	*Average abundance in the crust*[1] *(p.p.m.)*	$T_{\frac{1}{2}}^{*}$ *(years)*	*Sp. activity of av. crust matr. (γ/sec g)*
^{40}K	0.0118	2.5	$1.30 \cdot 10^{9}$	$6.7 \cdot 10^{-2}$
^{235}U	0.720	0.019	$7.1 \cdot 10^{8}$	$1.4 \cdot 10^{-3}$
^{176}Lu	2.60	0.02	$2.2 \cdot 10^{10}$	$6.9 \cdot 10^{-5}$
^{138}La	0.089	0.02	$1.1 \cdot 10^{11}$	$1.7 \cdot 10^{-5}$
^{48}Ca	0.185	26	$\sim 10^{14}$**	$\sim 10^{-5}$
^{50}V	0.25	0.34	$> 10^{16}$***	$< 10^{-9}$
^{130}Te	34.49	<0.017	$> 10^{18}$***	$< 10^{-12}$
^{58}Ni	67.76	<51	$> 10^{19}$***	$< 10^{-9}$

[1] From KRAUSKOPF (1967).
* When not specified, the reported values are from LEDERER et al. (1967).
** From DER METEOSIAN and GOLDHABER (1966).
*** From SONNTAG (1967).

emission by primeval radioisotopes in the crust. ^{232}Th and ^{238}U are primeval radioisotopes that decay to important secondary gamma emitters.

Secondary unstable isotopes are those whose half-lives are too short for them to occur unsupported in detectable amounts at the present time but which, nevertheless, are present in nature because they are continuously formed by natural decay processes or by natural nuclear reactions involving unstable primeval and stable isotopes. The highest contribution to the terrestrial gamma radiation field comes from radioisotopes generated through the sequence of decay transformations of three alpha emitting primeval radioisotopes (^{232}Th, ^{235}U, and ^{238}U). Each of them forms a decay chain, or radioactive series, ended by a stable isotope (in all three cases an isotope of Pb). Unstable members of the series are characterized by a half-life much shorter than that of the respective precursor and include isotopes of elements having atomic numbers between 81 and 92 (actually very minor amounts of the unstable isotope 206 of Hg are present in the ^{238}U series). The number (N_n) of nuclei of the nth member of a series which is present in a closed system, initially containing N^0 nuclei of precursor, after a time t is given by:

$$N_n(t) = N^0[C_1 \exp(-\lambda_1 t) + C_2 \exp(-\lambda_2 t) + \cdots + C_n \exp(-\lambda_n t)] \quad (4.1)$$

where:

$$C_1 = \frac{\lambda_1 \lambda_2 \ldots \lambda_{n-1}}{(\lambda_2 - \lambda_1)\ldots(\lambda_n - \lambda_1)} \qquad C_n = \frac{\lambda_1 \lambda_2 \ldots \lambda_{n-1}}{(\lambda_1 - \lambda_n)\ldots(\lambda_{n-1} - \lambda_n)}$$

and $\lambda_1 \lambda_2 \ldots \lambda_n$ = decay constants of the 1st (precursor), 2nd . . . nth member of the series.

When t is very large in comparison to $\lambda_2 \ldots \lambda_n$ an equilibrium condition is achieved where:

$$N_1 \lambda_1 = N_2 \lambda_2 = \cdots = N_n \lambda_n \quad (4.2)$$

i.e., the disintegration rate is the same for all the members of the series, the ratio of the number of nuclei of each member to that of any other member present in the system being inversely proportional to the ratio of the respective decay constants.

TABLE VII

ACTIVITIES OF SOME SPALLOGENIC RADIONUCLIDES IN RAINWATER COLLECTED IN THE AREA OF RICHLAND, WASHINGTON (U.S.A)
(After PERKINS et al., 1965)

Isotope	*Dis./min l*	*Collection time (min)*	*Date*
^{22}Na	0.19–0.36		June 15–18, 1964
^{24}Na	0.59–2.22		June 15–18, 1964
^{38}Cl	9–147	5–50	July 29–Aug. 18, 1964
^{39}Cl	~10–200	5–50	July 29–Aug. 18, 1964
^{38}S	4–13	5–50	July 29–Aug. 12, 1964

TABLE VIII

RADIOACTIVE PROPERTIES OF MAJOR GAMMA EMITTING SPALLOGENIC NUCLIDES

Isotope	*Half-life*	*Mode of decay*	*Gamma ray energy (MeV)*	*Absolute intensity (%)*	*Global inventory*[1]
$^{22}_{11}Na$	2.62 yr	β^+, EC	0.511	180	1.9 g
			1.275	100	
$^{24}_{11}Na$	14.96 h	β^-	1.369	100	
			2.754	100	
$^{38}_{17}Cl$	37.29 min	β^-	1.60	38	
			2.170	47	
$^{39}_{17}Cl$	55.5 min	β^-	0.246	44	
			1.27	50	
			1.52	42	
$^{38}_{16}S$	2.87 h	β^-	1.88	95	
$^{34m}_{17}Cl$	31.99 min	β^+, IT	0.145	45	
			0.511	100	
			1.17	12	
			2.12	38	
			3.30	12	
$^{7}_{4}Be$	53.6 days	EC	0.477	10.3	3.2 g
$^{26}_{13}Al$	$7.4 \cdot 10^5$ yr	β^+, EC	0.511	170	1.1 ton
			1.12	4	
			1.81	100	

[1] After LAL and PETERS (1967).

Very minor amounts of secondary gamma emitters are present in nature that have been originated by neutron induced reactions on ^{238}U, ^{235}U, and ^{232}Th (e.g., ^{237}Np, ^{233}U, ^{239}Pu*) or by spontaneous fission mostly of ^{238}U and ^{235}U, but their contribution to the total terrestrial gamma field is negligible.

Short lived nuclides are also continuously produced by cosmic ray induced reactions in the atmosphere. The main gamma emitters of this type are formed in the high atmosphere by spallation reactions with atmospheric Ar. PERKINS et al. (1965) report activities, observed in rainwater, of the main cosmic ray generated gamma emitters (Table VII). Table VIII lists their main radioactive characteristics after LEDERER et al. (1967).

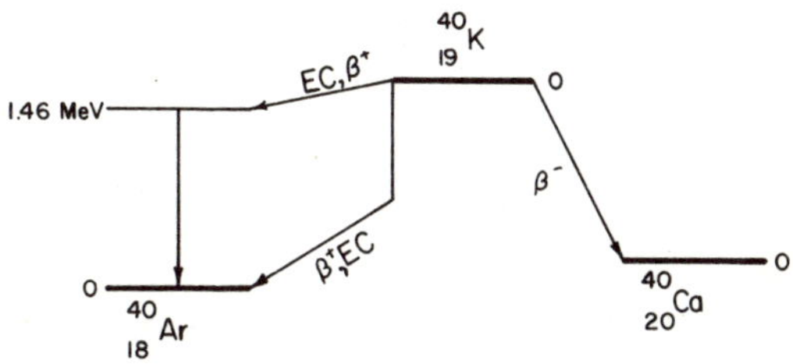

Fig.55. Decay scheme of ^{40}K.

Potassium 40

^{40}K is the only unstable isotope of potassium. Its natural isotopic abundance is 0.0119 ($\pm$0.0001)% according to NIER (1950), and 0.0118%, according to REUTERSWÄRD (1956); WHITE et al. (1956) and KENDALL (1960).

^{40}K undergoes branched decay by β^- emission to ^{40}Ca or mostly by electron capture to ^{40}Ar (Fig.55). Table IX reports various determinations of gamma specific activity of natural potassium, carried out after 1950. Table X summarizes the main radioactive properties of ^{40}K. The reported decay constants are those of widespread use in K/Ar geochronological work, i.e., the β^- decay constant given by KELLY et al. (1959) and by WETHERILL et al. (1956b) and the electron capture constant obtained by

* CHERDYNTZEV et al. (1965, 1967) reports finding ^{239}Pu in picogram amounts in Russian and Italian waters and volcanic rocks.

TABLE IX

DETERMINATIONS, CARRIED OUT AFTER 1950, OF THE SPECIFIC GAMMA ACTIVITY OF NATURAL K

	γ/sec g K		γ/sec g K		γ/sec g K
FAUST (1950)	3.6 (±0.4)	WASSERBURG and HAYDEN (1954)	3.66	Gerling et al. (1957)[1]	3.47
GRAF (1950)	3.4 (±0.5)	BACKENSTOSS and GOEBEL (1955)	3.50 (±0.14)	WETHERILL (1957)	3.39 (±0.12)
HOUTERMANS et al. (1950)	3.1 (±0.3)	Gerling et al. (1959)[1]	3.58	Gerling (1958)[1]	3.17
INGHRAM et al. (1950)	3.58	SUTTLE and LIBBY (1955)	2.96 (±0.3)	EGELKRAUT and LEUTZ (1960)	3.35 (±0.15)
SPIERS (1950)	2.97	MCNAIR et al. (1956)	3.33 (±0.15)	Saha and Gupta (1961)[1]	3.22 (±0.15)
BURCH (1953)	3.37 (±0.09)	WETHERILL et al. (1956a)	3.21	LEUTZ et al. (1965)	3.25 (±0.07)
		WETHERILL et al. (1956b)	3.24 (±0.15)		

[1] Reported by HOUTERMANS (1966).

TABLE X

RADIOACTIVE CONSTANTS OF ^{40}K

Decay mode	*Branching ratio*	*Half-life (years)*	*Decay constant (sec^{-1})*	*Radia-tion*	*Energy (MeV)*	*Absolute intensity (%)*
β^- emission	0.89	$1.47 \cdot 10^9$	$1.50 \cdot 10^{-17}$	β^-	1.314 (max)	89
Electron capture	0.11	$1.18 \cdot 10^{10}$	$0.186 \cdot 10^{-17}$	γ	1.461	11

Total decay constant $= \lambda = \lambda_\beta + \lambda_{ec} = 1.69 \cdot 10^{-17}$ sec^{-1}.
Total half-life $= 1.30 \cdot 10^9$ yr.

WETHERILL (1957) by scintillation counting. Different "best values" have been chosen by other authors. HOUTERMANS (1966) chooses as best constants those obtained by WETHERILL et al. (1956*a*) by geological methods, i.e., $\lambda_\varepsilon = 0.183 \cdot 10^{-17}$ sec^{-1}, $\lambda_{\beta^-} = 1.50 \cdot 10^{-17}$ sec^{-1}, leading to a total half-life of $T_{\frac{1}{2}} = 1.31$ gigayears and to a branching ratio $\lambda_\varepsilon/\lambda_{\beta^-} = 11.7(\pm 1.5)\%$. LEDERER et al. (1967) report as the most accurate half-life $T_{\frac{1}{2}} = 1.26$ gigayears based upon $\lambda_{\beta^-} = 1.55 \cdot 10^{-17}$ sec^{-1} (GLENDENIN, 1961, in satisfactory agreement with LEUTZ et al., 1965) and $\lambda_{\beta^-}/\lambda_\varepsilon + \lambda_{\beta^-} = 0.89$.

β^- emission occurs when the ^{40}K decays directly to ^{40}Ca in the ground state, whereas electron capture leaves in most cases the ^{40}Ar nucleus in a 1.46 MeV excited state, and is followed by a prompt 1.46 MeV gamma emission. The absolute intensity of the electron capture decay directly to the ground state, and therefore with no gamma emission, has been evaluated to be about 0.2% and the positron emission 0.0013% (LEUTZ et al., 1965). Electron capture decay in each case is followed by the emission of X-rays characteristic of Ar (K X-rays energy = 3 keV; fluorescence yield = 0.81). AZMAN et al. (1968) have evaluated the yield for K/L + M + ... ratio for electron capture to be 0.44 ($\pm$0.09).

Thorium 232 series

Table XI lists in sequence the members of the ^{232}Th decay series, their half-lives ($T_{\frac{1}{2}}$), decay constants (λ) and mode of decay. Where not specified, all the radioactive properties reported throughout this chapter are from LEDERER et al. (1967). Each isotope is reported

TABLE XI

^{232}Th SERIES

Isotope	*Classical name*	$T_{\frac{1}{2}}$	$\lambda(sec^{-1})$	*Primary decay mode*	*Remarks*
$^{232}_{90}Th$	Thorium	$1.41 \cdot 10^{10}$ yr	$1.56 \cdot 10^{-18}$	α	
$^{228}_{88}Ra$	Mesothorium 1	6.7 yr	$3.28 \cdot 10^{-9}$	β^-	
$^{228}_{89}Ac$	Mesothorium 2	6.13 h	$3.14 \cdot 10^{-5}$	β^-	
$^{228}_{90}Th$	Radiothorium	1.910 yr	$1.15 \cdot 10^{-8}$	α	
$^{224}_{88}Ra$	Thorium X	3.64 days	$2.21 \cdot 10^{-6}$	α	
$^{220}_{86}Rn$	Thoron	55.3 sec	$1.25 \cdot 10^{-2}$	α	
$^{216}_{84}Po$	Thorium A	0.145 sec	4.78	α	
$^{212}_{82}Pb$	Thorium B	10.64 h	$1.81 \cdot 10^{-5}$	β^-	
$^{212}_{83}Bi$	Thorium C	60.60 min	$1.91 \cdot 10^{-4}$	α, β^-	branched decay: β^- (66.3%) to ^{212}Po α(33.7%) to ^{208}Tl
$^{212}_{84}Po$	Thorium C′	$3.04 \cdot 10^{-7}$ sec	$2.28 \cdot 10^{6}$	α	
$^{208}_{81}Tl$	Thorium C″	3.10 min	$3.73 \cdot 10^{-3}$	β^-	
$^{208}_{82}Pb$	Thorium D	stable			

with its standard isotopic designation as well as with its classical name. However, the standard isotopic designation will be preferred in this discussion because, being a precise chemical identification of the element, it is believed to be more useful from a geochemical point of view.

The comparatively short half-life of all the decay products of ^{232}Th allows a relatively quick achievement of equilibrium in natural systems. Fig.56 shows the recovery curve of the longest lived daughter, ^{228}Ra, up to equilibrium with ^{232}Th; this equilibrium is practically achieved after only about 40 years. Because the second longest lived daughter, ^{228}Th, re-establishes equilibrium in about 5 years with ^{228}Ra, 45 years can be considered the maximum time for re-establishing secular radioactive equilibrium throughout the series in a closed system. The influence of episodic opening of the system for any member is therefore quickly cancelled and equilibrium conditions can be thought to prevail in natural systems. Disequilibrium conditions between ^{228}Th and ^{232}Th have been observed in sea water (MOORE and SACKETT,

1964) probably as a result of continuous leaching of ^{228}Ra from sediments and/or longer residence time of ^{228}Ra in respect to ^{232}Th in sea water.

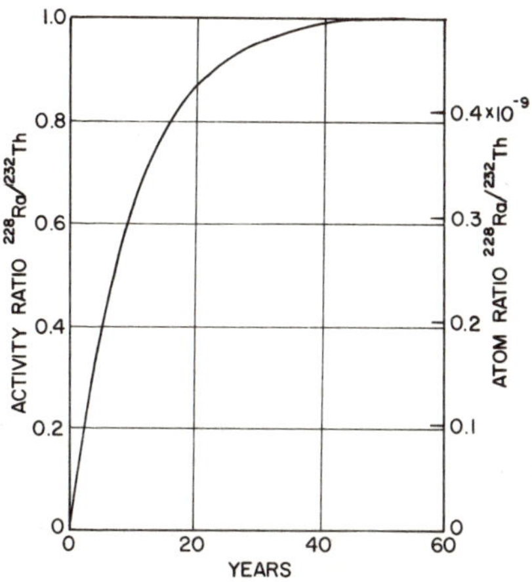

Fig.56. Recovery curve of ^{228}Ra up to equilibrium with ^{232}Th. (After HYDE et al., 1964; courtesy of Prentice Hall Publ. Co.)

A minor fraction of the gaseous member of the series, ^{220}Rn, diffuses continuously out of the earth's crust into the atmosphere, where it is present with all its decay products. It should be noted that, because the achievement of equilibrium is only possible when the parent isotope is longer lived than its decay product, unsupported ^{220}Rn ($T_{\frac{1}{2}}$ = 55.3 sec) cannot reach equilibrium with its β^- and gamma emitting daughters because of the presence of longer lived ^{212}Pb ($T_{\frac{1}{2}}$ = 10.6 h). Atmospheric concentrations of ^{220}Rn evaluated upon β^- or gamma activity of its decay products are therefore lower than true, as pointed out by ISRÄEL (1964a,b); ISRÄEL and ISRÄEL (1965).

Fig.57 reports the simplified decay schemes of the whole ^{232}Th series and Table XII lists the energy and absolute intensities of particles and radiations emitted from the members of this series. Only transitions having higher than 1% absolute intensity have been reported here; the main gamma transitions lower than 1% absolute intensities have been reported by dotted lines in the decay

TABLE XII

DETAILED DECAY PROPERTIES OF THE ^{232}Th SERIES

Nuclide	α energy (MeV)	β^- max. en. (MeV)	Absolute intensity (%)	Gamma en. (MeV)	Absolute intensity (%)	Conv. electrons abs. int. (%)	Conversion X-ray	Energy (keV)
$^{232}_{90}$Th	3.95		24	0.059	–	24	$_{88}$Ra–L	~18
	4.01		76					
$^{228}_{88}$Ra		0.024	30	0.010		30	$_{87}$Ac–M	~5
		0.040	70					
$^{228}_{89}$Ac		0.45	13	0.058	–	53	$_{90}$Th–L	~20
		0.64	7·6	0.129	4.1	5.2	$_{90}$Th–L	~20
		1.11	53	0.209	1			
		1.70	6.7	0.270	3			
		1.85	9.6	0.328	4			
		2.18	10.1	0.338	11			
				0.908	25			
				0.960 + 0.966	20			
$^{228}_{90}$Th	5.34		28	0.084	1.6	~25	$_{88}$Ra–L	~18
	5.43		71					
$^{224}_{88}$Ra	5.45		6	0.241	3.7	1.3	$_{86}$Rn–K–L	~84 and ~17
	5.68		94					

TABLE XII (*continued*)

Nuclide	*α energy (MeV)*	*β^- max. en. (MeV)*	*Absolute intensity (%)*	*Gamma en. (MeV)*	*Absolute intensity (%)*	*Conv. electrons abs. int. (%)*	*Conversion X-ray*	*Energy (keV)*
$^{220}_{86}Rn$	6.29		100					
$^{216}_{84}Po$	6.78		100					
$^{212}_{82}Pb$		0.154	5	0.115		2.1	$_{83}$Bi–K	~77
		0.331	81	0.239	47	28	$_{83}$Bi–K–	~77
		0.569	14	0.300	3.2	1	$_{83}$Bi–K	~77
$^{212}_{83}Bi$	6.05		23.5	0.040	2	46	$_{81}$Tl–L	~15
	6.09		9	0.727	7.1			
		0.67	2.2	0.785	1.1			
		0.93	1.4	1.620	1.8			
		1.55	5					
		2.27	54					
$^{212}_{84}Po$	8.78		100					
$^{208}_{81}Tl$		1.04	4.6	0.277	7	2.3	$_{82}$Pb–K	~72
		1.29	24	0.511	23	1.6	$_{82}$Pb–K	~72
		1.52	23	0.583	86	1.3	$_{82}$Pb–K	~72
		1.80	49	0.763	2			
				0.860	12			
				2.615	100			

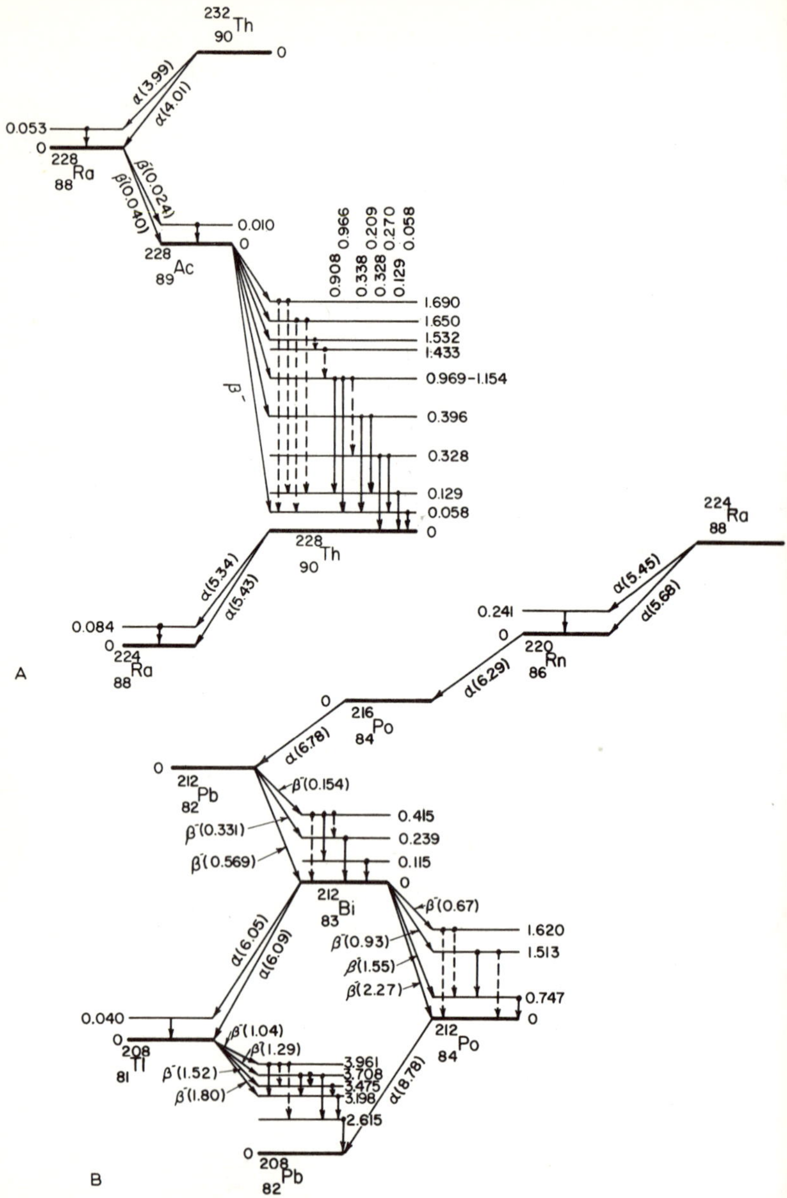

Fig.57. Simplified decay schemes of the members of the ^{232}Th series.

schemes. Table XII reports also the main conversion X-rays; fluorescence yields are approximately 0.96 for all the main K X-rays emitters of the series (Fig.5).

A typical gamma spectrum of the whole ^{232}Th series in equilibrium, obtained by NaI(Tl) scintillation spectrometry, is reported in Appendix 2. The most prominent features of the spectrum are full energy peaks at 2.615 MeV (^{208}Tl), around 0.95 MeV (0.910 + 0.966 + 0.960 from ^{228}Ac), 0.58 MeV (0.583 + 0.511 from ^{208}Tl), 0.33 MeV (0.30 from ^{212}Pb + 0.33 + 0.34 from ^{228}Ac), 0.24 MeV (0.24 from ^{212}Pb + 0.27 from ^{228}Ac + 0.28 from ^{208}Tl), and 0.09 MeV (^{228}Th and conversion K X-ray from Rn). Lower full energy peaks occur at 0.86 MeV (^{208}Tl) and at 0.75 MeV (^{208}Tl and ^{212}Bi).

Uranium 238 *series*

Table XIII lists in sequence the members of the ^{238}U series, their half-life ($T_{\frac{1}{2}}$), decay constant (λ) and mode of decay. It must be pointed out that constants largely different from the chosen ones have been reported for ^{230}Th and ^{210}Pb. The decay constant reported in Table XIII for ^{230}Th is based upon specific activity measurements (HYDE, 1964); ATTREE et al. (1962) report 7.52 $(\pm 0.16) \cdot 10^4$ years upon calorimetric measurements. For ^{210}Pb the decay constant given by RAMTHUN (1964), in reasonable agreement with the value found by ECKELMANN et al. (1960) by the geological method, has been reported. Determinations of this decay constant by dual ion chamber techniques have yielded a half-life of 19.4 years (TOBAILEM, 1955) and 20.4 years (HARBOTTLE, 1959).

The presence of long lived ^{234}U, ^{230}Th, and ^{226}Ra makes the establishment of secular radioactive equilibrium throughout the series a long process. Furthermore, the different chemical natures of the long lived intermediate members enhance the chances of chemical fractionation in nature and, in this case, an episodical removal or support of any long lived member produces long term disequilibria conditions.

Intermediate long lived members split the ^{238}U series into two parts, in which equilibrium is reached in a comparatively short time. The part preceeding the long lived intermediate members includes only ^{238}U, ^{234}Th, and ^{234}Pa. The longest lived decay product,

TABLE XIII

^{238}U SERIES

Isotope	*Classical name*	$T_{\frac{1}{2}}$	$\lambda(sec^{-1})$	*Primary decay mode*	*Remarks*
$^{238}_{92}U$	Uranium I	$4.51 \cdot 10^9$ yr	$4.88 \cdot 10^{-18}$	α	
$^{234}_{90}Th$	Uranium X_1	24.10 days	$3.33 \cdot 10^{-7}$	β^-	
$^{234m}_{91}Pa$	Uranium X_2	1.175 min	$9.83 \cdot 10^{-3}$	β^-	isomeric transition to ^{234}Pa (0.13%)
$^{234}_{91}Pa$	Uranium Z	6.75 h	$2.85 \cdot 10^{-5}$	β^-	
$^{234}_{92}U$	Uranium II	$2.47 \cdot 10^5$ yr	$8.91 \cdot 10^{-14}$	α	
$^{230}_{90}Th$	Ionium	$8.0 \cdot 10^4$ yr	$2.75 \cdot 10^{-13}$	α	
$^{226}_{88}Ra$	Radium	1,602 yr	$1.37 \cdot 10^{-11}$	α	
$^{222}_{86}Rn$	Radon	3.8223 days	$2.10 \cdot 10^{-6}$	α	
$^{218}_{84}Po$	Radium A	3.05 min	$3.79 \cdot 10^{-3}$	α, β^-	branched decay: α to ^{214}Pb (99.98%); β^- to ^{218}At (0.02%)
$^{214}_{82}Pb$	Radium B	26.8 min	$4.31 \cdot 10^{-4}$	β^-	
$^{218}_{85}At$	Astatine-218	~2 sec	~0.35	α	
$^{214}_{83}Bi$	Radium C	19.7 min	$5.86 \cdot 10^{-4}$	α, β^-	branched decay: α to ^{210}Tl (0.04%); β^- to ^{214}Po (99.96%)
$^{214}_{84}Po$	Radium C′	$1.64 \cdot 10^{-4}$ sec	$4.23 \cdot 10^3$	α	
$^{210}_{81}Tl$	Radium C″	1.32 min	$8.75 \cdot 10^{-2}$	β^-	
$^{210}_{82}Pb$	Radium D	22.0 yr	$1 \cdot 10^{-9}$	β^-	branched decay: α to ^{206}Hg ($1.8 \cdot 10^{-6}$%) β^- to ^{210}Bi (~100%)
$^{210}_{83}Bi$	Radium E	5.013 days	$1.60 \cdot 10^{-6}$	β^-	branched decay: α to ^{206}Tl (~10^{-5}%) β^- to ^{210}Po (~100%)
$^{210}_{84}Po$	Radium F	138.4 days	$5.78 \cdot 10^{-8}$	α	
$^{206}_{82}Pb$	Radium G	stable			

^{234}Th, reaches equilibrium with ^{238}U after about 200 days (Fig.58). As discussed in Chapter 1, this part includes the only case of isomerism occurring in nature, ^{234m}Pa. Fig.59 and Table XIV report simplified decay schemes and emission characteristics of the members of this part of the series. Isomeric transition from ^{234m}Pa to ^{234}Pa and consequent β^- decay of ^{234}Pa to ^{234}U have

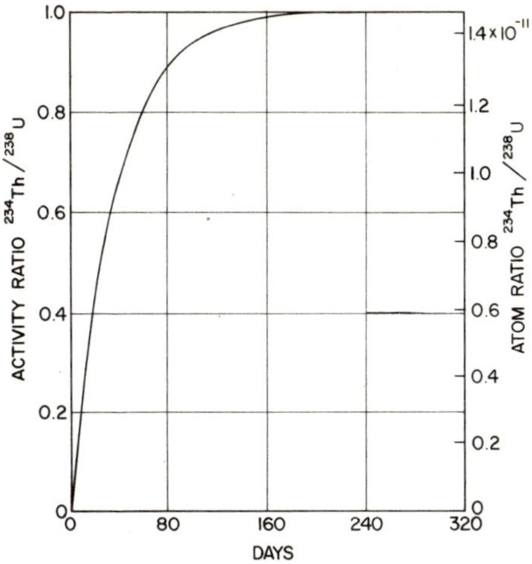

Fig.58. Recovery curve of ^{234}Th up to equilibrium with ^{238}U. (After HYDE et al., 1964; courtesy of Prentice Hall Publ. Co.)

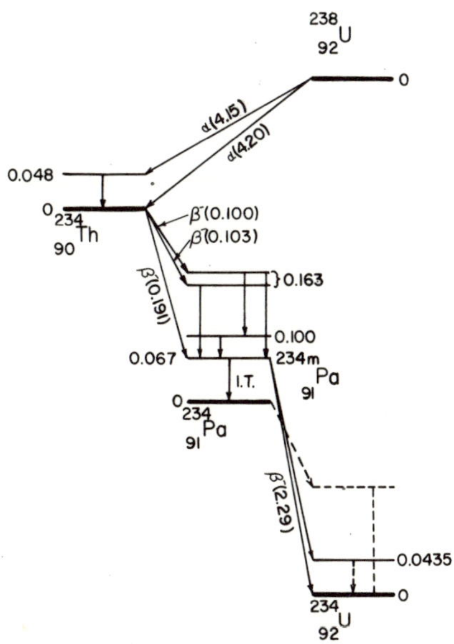

Fig.59. Simplified decay schemes of ^{238}U, ^{234}Th, ^{234m}Pa, ^{234}Pa.

TABLE XIV

DETAILED DECAY PROPERTIES OF PRE-^{226}Ra MEMBERS OF ^{238}U SERIES

Nuclide	*α energy (MeV)*	*β⁻ max. en. (MeV)*	*Absolute intensity (%)*	*Excited levels (MeV)*	*Gamma abs. int. (%)*	*Conv. electrons abs. int. (%)*	*Conversion X-rays*	*Energy (keV)*
$^{238}_{92}U$	4.15		23	0.048		23	$_{90}$Th–L	~20
	4.20		77					
$^{234}_{90}Th$		0.100*	7*	0.093	4	8*	$_{91}$Pa–L	~20
		0.103	12	0.064	3.5	1.1	$_{91}$Pa–L	~20
		0.191	81	0.030		4.5	$_{91}$Pa–M	5
$^{234m}_{91}Pa$	2.29	98%						
$^{234}_{92}U$	4.72		28	0.053		~26	$_{90}$Th–L	~20
	4.77		72					
$^{230}_{90}Th$	4.62		24	0.068		~24	$_{88}$Ra–L	~18
	4.69		76					

* After HYDE et al. (1964).

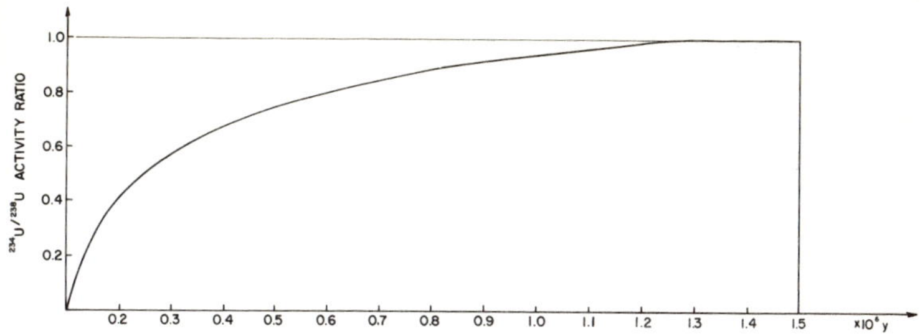

Fig.60. Recovery curve of ^{234}U up to equilibrium with ^{238}U.

been reported by dotted lines in the decay scheme, because the absolute intensity of the transition is less than 1% and the relative emission characteristics have been omitted in Table XIV.

This part of the series has a very weak gamma activity; practically the only significant gamma emissions are low energy photons (0.06 and 0.09 MeV) from ^{234}Th. Conversion X-rays peak around 20 (Th–L and Pa–L) and around 5 (Pa–M) keV.

^{224m}Pa undergoes β^- decay to ^{234}U which, together with its descendants ^{230}Th and ^{226}Ra, form the long lived intermediate part of the series. The respective recovery curves show that ^{234}U

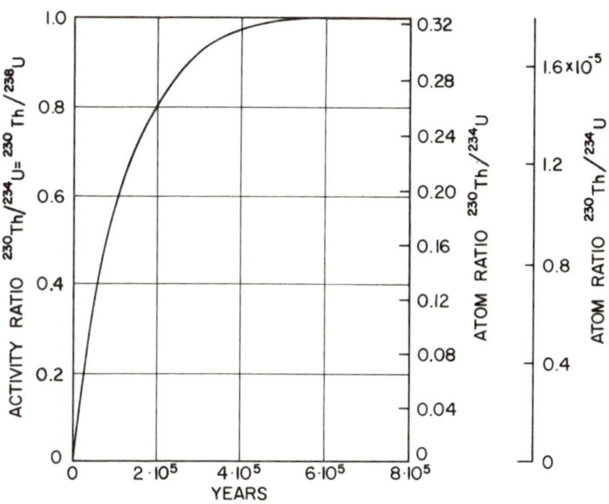

Fig.61. Recovery curve of ^{230}Th up to equilibrium with ^{234}U–^{238}U. (After HYDE et al., 1964; courtesy of Prentice Hall Publ. Co.)

reaches equilibrium with ^{238}U (Fig.60) after about 10^6 years, and ^{230}Th reaches equilibrium with ^{234}U–^{238}U after about $4 \cdot 10^5$ years (Fig.61). ^{226}Ra reaches equilibrium with ^{230}Th after $1 \cdot 10^4$ years (Fig.62).

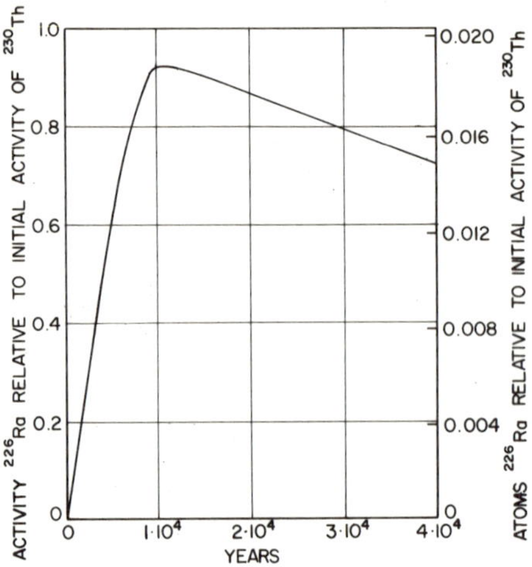

Fig.62. Recovery curve of ^{226}Ra up to equilibrium with unsupported ^{230}Th. (After HYDE et al., 1964; courtesy of Prentice Hall Publ. Co.)

Wide fluctuations of ^{234}U, ^{230}Th, and ^{226}Ra from equilibrium values have been observed in nature. Preferential leaching of ^{234}U in respect of ^{238}U from rocks causes a depletion of ^{234}U in weathered rocks and soils, and an average enrichment of ^{234}U of about 15% above the equilibrium content in fresh waters (CHAVLOV, 1959; CHERDYNTZEV et al., 1961; THURBER, 1962; RICHARDSON, 1963, 1964; ROSHOLT et al., 1963, 1964; SYROMYATNIKOV et al., 1967). Differential precipitation from waters produces enrichments of ^{230}Th in respect to both U isotopes in sediments, as first hypothesized by PETTERSON (1937) and confirmed by KOCZY et al. (1967), SACKETT et al. (1958) and many other researchers. By contrast ^{230}Th is depleted in fossil molluscs, oolites and corals (for details, see the discussions by WETHERILL and TILTON, 1967; and ROSHOLT, 1967). Depletion of ^{230}Th has been observed also in stalactites and stalagmites (FORNACA-RINALDI, 1968a,b). ^{230}Th

variations from the equilibrium contents have been observed also in young igneous rocks (SOMAYAJULU et al., 1966; KIGOSHI, 1967; CHERDYNTZEV et al., 1967; TADDEUCCI et al., 1967; OVERSBY and GAST, 1968). Unsupported amounts of ^{226}Ra have been reported in volcanics from Vesuvius (CASERTANO, 1952; IMBO', 1955; FERRETTI SFORZINI et al., 1955; VITTOZZI and GASPARINI, 1964; OVERSBY and GAST, 1968), and excesses of ^{226}Ra above the equilibrium value has been reported for a freshly erupted lava from Stromboli (GASPARINI et al., 1967). No significant gamma emission occurs from ^{234}U and ^{230}Th; only ^{226}Ra emits a weak 0.186 MeV gamma (Figs.63, 64, and Tables XIV, XV).

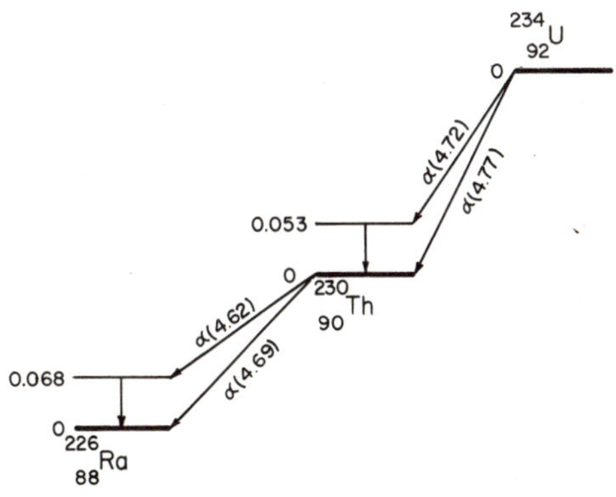

Fig.63. Decay schemes of ^{234}U and ^{230}Th.

All major gamma emitters of the series are in the part following the long lived members. The longer lived member of this final part of the series is ^{210}Pb ($T_{\frac{1}{2}}$ = 22 years). Complete equilibrium from ^{226}Ra to ^{214}Po is reached after about 32 days (Fig.65), whereas ^{210}Pb and its descendants reach equilibrium with ^{226}Ra after about 100 years (Fig.66).

^{222}Rn, the gaseous member of the series, diffuses continuously out of the rocks and soils into the atmosphere, where it is present with its decay products. Equilibrium from ^{222}Rn to ^{214}Po is promptly achieved but longer lived ^{210}Pb is present in excessive amounts. Rain and snow remove ^{210}Pb from the atmosphere and it is found trapped and unsupported in glaciers, and can be used for dating purposes (GOLDBERG, 1963; CROZAZ et al., 1964; CROZAZ, 1967). Unsupported ^{210}Pb has been found also in volcanic sublimates (EBERHART et al., 1955; HOUTERMANS et al., 1964). SOMAYAJULU et al. (1966) have reported wide fluctuations in the ^{210}Pb content of lavas.

The most active gamma emitters (Table XV, Fig.64) are ^{214}Bi and ^{214}Pb; weaker gamma emitters are ^{226}Ra and ^{210}Pb. Three branched decays occur in this part of the series, i.e., ^{214}Bi (β^- to ^{214}Po or α to ^{210}Tl), ^{210}Pb (β^- to ^{210}Bi or α to ^{206}Hg), and ^{210}Bi (β^- to ^{210}Po or α to ^{206}Tl). In any case, the transitions to ^{210}Tl (0.04%), to ^{206}Hg ($1.8 \cdot 10^{-6}$%) and to ^{206}Tl ($\sim 10^{-5}$%) are negligible.

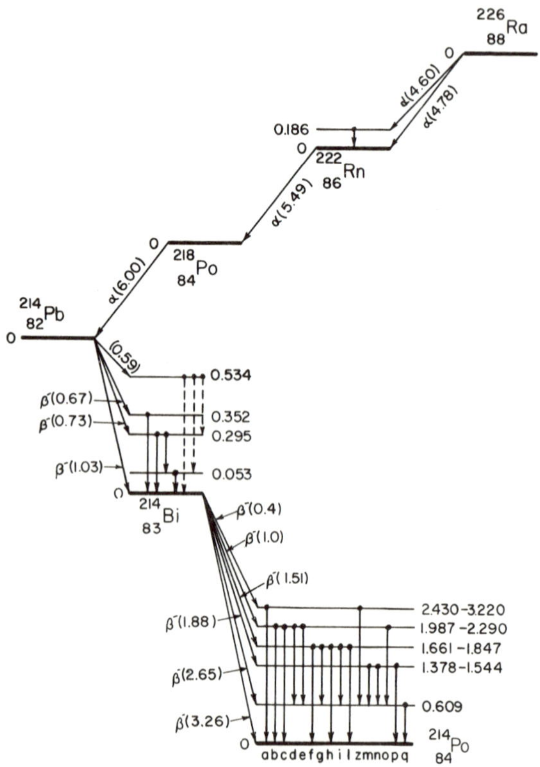

A

a = 2.44; b = 2.20; c = 2.12; d = 1.41; e = 1.24; f = 1.76; g = 1.16; h = 1.73; i = 1.12; l = 1.66; z = 1.85; m = 0.935; n = 0.769; o = 1.28; p = 0.609; q = 1.51

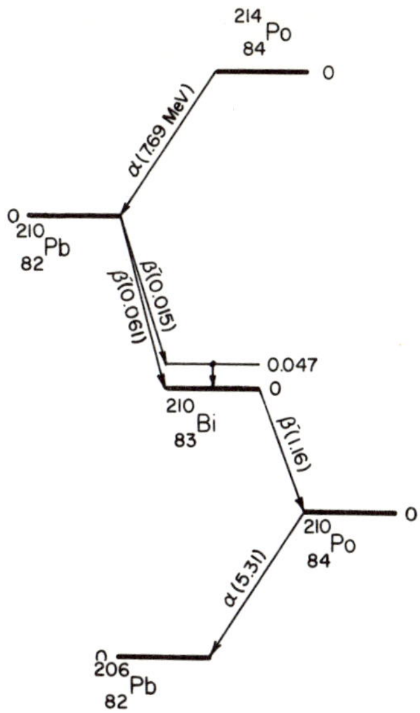

B

Fig.64. Simplified decay schemes of the post ^{226}Ra members of the ^{238}U series. All the branched decays whose probability of occurrence is less than 1% have been omitted.

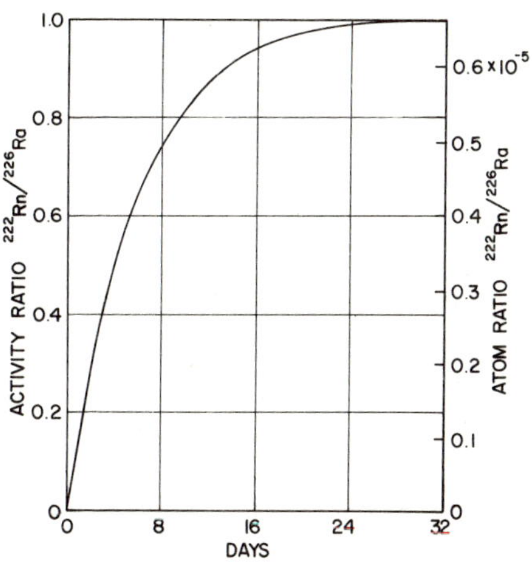

Fig.65. Recovery curve of ^{222}Rn up to equilibrium with ^{226}Ra. (After HYDE et al., 1964; courtesy of Prentice Hall Publ. Co.)

A typical NaI(Tl) scintillation spectrum of the whole ^{238}U series in secular equilibrium is shown in Appendix 2. There is only one full energy peak (0.09 MeV) where gamma rays from the ^{232}U, ^{234}Th, ^{234}Pa group contribute significally. All the full energy peaks above 0.09 MeV are mostly due to ^{226}Ra and its radioactive daughters. The 0.19 MeV full energy peak is due to ^{226}Ra (and ^{235}U); 0.25, 0.30 and 0.35 MeV are mostly due to ^{214}Pb and all the full energy peaks above 0.35 MeV to ^{214}Bi.

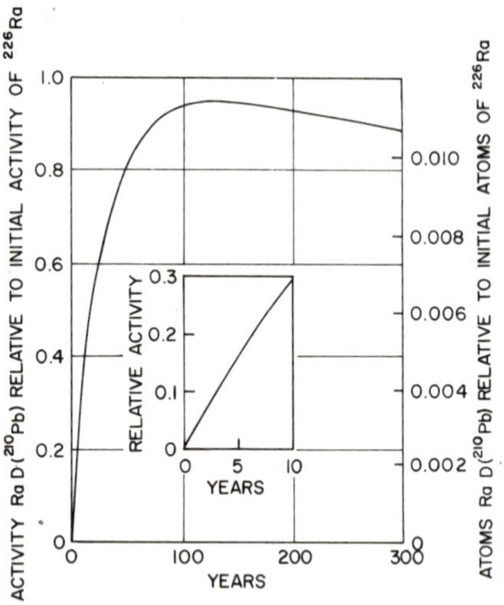

Fig.66. Recovery curve of ^{210}Pb up to unsupported ^{226}Ra. (After HYDE et al., 1964; courtesy of Prentice Hall Publ. Co.)

Uranium 235 *series*

Table XVI reports the members, half-lives, decay constants and modes of decay of the ^{235}U series. The decay constant of ^{235}U reported in Table XVI is that currently accepted and has a precision of $\pm 2\%$ (FLEMING et al., 1952). BANKS and SILVER (1966) have evaluated a decay constant of $3.10 \cdot 10^{-17}$ sec^{-1}, upon lead isotopic ratios in cogenetic zircons and uranothorites. The ^{235}U natural isotopic ratio is 0.7114($\pm$0.0007)% (HYDE et al., 1964) and this ratio has not been found to vary in nature; the disintegration rate of ^{235}U in natural U is only 4.6% of that of ^{238}U. The activity

DETAILED DECAY PROPERTIES OF ^{226}Ra AND DAUGHTERS

Nuclide	*α energy (MeV)*	*β⁻ max. en. (MeV)*	*Absolute intensity (%)*	*Gamma en. (MeV)*	*Absolute intensity (%)*	*Conv. electrons abs. int. (%)*	*Conversion X-rays*	*Energy (keV)*
$^{226}_{88}$Ra	4.60		5.4	0.186	4	~1.4	mostly L–Rn	~17
	4.78		94.6					
$^{222}_{86}$Rn	5.49		100					
$^{218}_{84}$Po	6.00		~100					
$^{214}_{82}$Pb		0.59	~2	0.053	~1			
		0.67	47	0.242	4	~3	mostly $_{83}$Bi–K	~80
		0.73	44	0.295	19		mostly $_{83}$Bi–K	
		1.03	6	0.352	36	11	mostly $_{83}$Bi–K	~80
$^{214}_{83}$Bi		0.4	9	0.609	47			
		1.0	23	0.665	2.3			
		1.51	40	0.769	5.3			
		1.88	9	0.787	1.2			
		3.26	19	0.805	1.5			
				0.935	3.3			
				1.12	16			
				1.155	1.8			
				1.26	7.7			
				1.40	8.8			
				1.51	2.4			
				1.66	1.2			
				1.73	3.2			
				1.76	17			
				1.85	2.3			
				2.12	1.3			
				2.20	6			
				2.44	2			
$^{214}_{84}$Po	7.69		100					
$^{210}_{82}$Pb		0.015	81	0.047	4.1	~66	$_{83}$Bi–L	~15
		0.016	19					
$^{210}_{83}$Bi		1.16	100					
$^{210}_{84}$Po	5.31		100					

TABLE XVI

^{235}U SERIES

Isotope	*Classical name*	$T_{\frac{1}{2}}$	λ (*sec*$^{-1}$)	*Primary decay mode*	*Remarks*
$^{235}_{92}U$	Actiniouranium	$7.1 \cdot 10^{8}$ yr	$3.09 \cdot 10^{-17}$	α	
$^{231}_{90}Th$	Uranium Y	25.52 h	$3.70 \cdot 10^{-4}$	β^-	
$^{231}_{91}Pa$	Protoactinium	$3.48 \cdot 10^{4}$ yr*	$6.32 \cdot 10^{-13}$	α	
$^{227}_{89}Ac$	Actinium	21.6 yr	$1 \cdot 10^{-9}$	α, β^-	branched decay: 98.8% β^- to ^{227}Th, 1.2% α to $^{223}F_2$
$^{227}_{90}Th$	Radioactinium	18.2 days	$4.41 \cdot 10^{-7}$	α	
$^{223}_{87}Fr$	Actinium K	22 min	$5.25 \cdot 10^{-4}$	β^-	
$^{223}_{88}Ra$	Actinium X	11.435 days	$7.88 \cdot 10^{-6}$	α	
$^{219}_{86}Rn$	Actinon	4.00 sec	0.173	α	
$^{215}_{84}Po$	Actinium A	$1.778 \cdot 10^{-3}$ sec	$3.90 \cdot 10^{2}$	α	
$^{211}_{82}Pb$	Actinium B	36.1 min	$3.20 \cdot 10^{-4}$	β^-	
$^{211}_{83}Bi$	Actinium C	2.16 min	$5.33 \cdot 10^{-3}$	α, β^-	branched decay: 0.32% β^- to ^{211}Po, 99.68% α to ^{207}Tl
$^{211}_{84}Po$	Actinium C′	0.52 sec	0.13	α	
$^{207}_{81}Tl$	Actinium C″	4.79 min	$2.41 \cdot 10^{-3}$	β^-	
$^{207}_{82}Pb$	Actinium D	Stable			

$^{223}_{87}Fr$ undergoes also $4 \cdot 10^{-3}$% α decay to 0.9 min $^{219}_{85}At$, which, in its turn undergoes branched decay: 3% β^- to ^{219}Rn and 97% α to 8 min ^{215}Bi, which β^- decays to ^{215}Po.

* After HYDE et al. (1964).

of all the members of the ^{235}U series is therefore overshadowed by the much more prominent activity of the ^{238}U series and, in natural systems, can be neglected for all practical gamma spectrometric purposes.

The only long lived member of the series is ^{231}Pa; it reaches equilibrium with ^{235}U after about $2 \cdot 10^{5}$ years (Fig.67). Disequilibrium amounts of ^{231}Pa have been reported in marine fossils (DOOLEY et al. 1964; ROSHOLT, 1967; SAKANQUE et al., 1967).

Table XVII reports the characteristics of the main particles and radiation emitted from all the members of this series. The series has a gamma activity that is concentrated in the range 0.026–0.351 MeV.

TABLE XVII

DETAILED DECAY CONSTANTS OF ^{235}U SERIES

Nuclide	*α energy (MeV)*	*β⁻ max. en. (MeV)*	*Absolute intensity (%)*	*Gamma en. (MeV)*	*Absolute intensity (%)*	*Conv. electrons abs. int. (%)*	*Conversion X-rays*	*Energy*
$^{235}_{92}U$	4.40		84*	0.143	11	7	$_{90}Th$–K	
	4.55		10.5	0.185	54			
$^{231}_{90}Th$		0.134	20	0.026	12	73	$_{91}Pa$–M	
		0.218	33	0.059		74	$_{91}Pa$–M	
		0.299	39	0.082		41	$_{91}Pa$–M	
$^{231}_{91}Pa$	4.73		11	0.290*	6			
	4.95		22					
	5.02*		47					
	5.06		10					
$^{227}_{89}Ac$		0.046*	89					
$^{227}_{90}Th$	5.71*		14	0.237	15			
	5.76		21					
	5.98		24					
	6.04		23					
$^{223}_{88}Ra$	5.57*		35	0.147*	12	25	$_{86}Rn$–K	
	5.73		63	0.270	10			
$^{219}_{86}Rn$	6.42		8					
	6.55		11					
	6.82		81					
$^{215}_{84}Po$	7.38		100					
$^{211}_{82}Pb$		1.36	92					
$^{211}_{83}Bi$	6.28		16	0.351	14			
	6.62		84					
$^{211}_{84}Po$	7.45		99					

* After HYDE et al. (1964).

In equilibrium conditions, the most prominent feature is the 0.185 MeV full energy peak due to ^{235}U. Reported in Appendix 2 is a gamma spectrum of natural U which has not yet grown up detectable post ^{234}U gamma emitting daughters. Full energy peaks occur at 0.09 MeV (^{234}Th), 0.14 and 0.19 (^{235}U) MeV. Minor peaks can be discerned at 0.29 MeV (^{235}U sum peak and ^{231}Pa), 0.71 and 1.05 MeV (^{234}Pa and ^{234m}Pa). For a more detailed description of the radioactive properties of the three described natural series the reader is referred to the extensive work by HYDE et al. (1964).

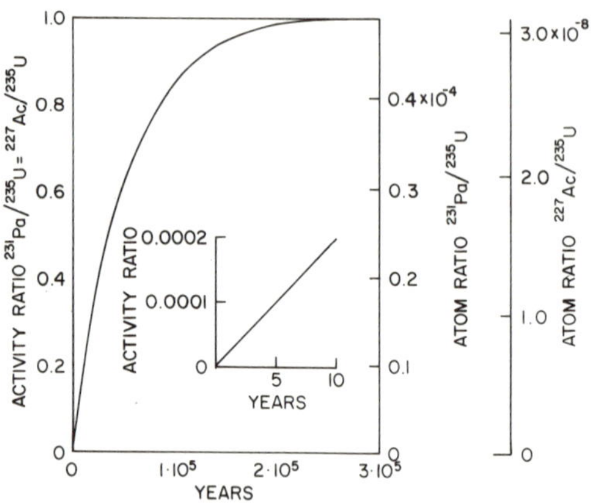

Fig.67. Recovery curve of ^{231}Pa up to equilibrium with ^{235}U. (After HYDE et al., 1964; courtesy of Prentice Hall Publ. Co.)

ARTIFICIAL EMITTERS

Since the early nineteen forties, when fission energy had its first applications, a large number of newly formed unstable isotopes has been introduced into the earth's natural environment. Most of these have been introduced into the atmosphere after atmospheric nuclear explosions and are subsequently transferred to the surface soils and waters both by settling through the air and by rainwaters. Reactors also contribute their radioactive wastes as well as impurities contained in coolant waters.

Fig.68 reports the fission yields (i.e., the percentage of fission events giving rise to a given nuclide) of different mass numbers for thermal neutron induced fission of ^{233}U, ^{235}U, and ^{239}Pu (after HYDE, 1964). Fission products are practically bracketed between mass numbers 70 and 170 with peak values in the regions 90–100

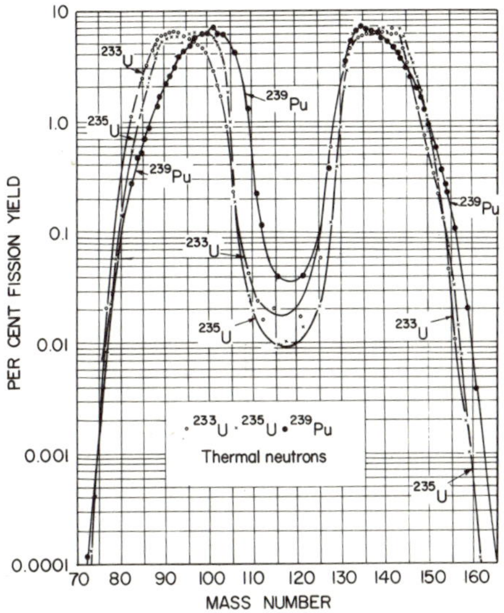

Fig.68. Fission yields of different mass numbers isotopes for thermal neutrons induced fission of ^{233}U, ^{235}U, and ^{239}Pu. (After HYDE et al., 1964; courtesy of Prentice Hall Publ. Co.)

and 135–140. The position of the peaks shows that the fission is asymmetrical, i.e., the fissioned nucleus splits into two fragments having unequal numbers of nucleons. Most fission products are unstable and, having a high excess of neutrons, decay by β^- emission forming short decay chain. Unstable fission products can be subdivided into short lived (half-life not longer than a few days), intermediate (half-life in the range from some days to some years), and long lived (half-life longer than a few years). Tables XVIII, XIX, and XX report the main radioactive characteristics of gamma emitting fission products in order of increasing half-life. Only isotopes having half-life longer than 1 h have been reported; when

TABLE XVIII

DECAY CONSTANTS AND FISSION YIELDS (F.y.) OF THE MAIN SHORT-LIVED FISSION PRODUCTS

Fission product	*Decay mode*	*Half-life*	*F.y. for* ^{233}U (%)	*F.y. for* ^{235}U (%)	*F.y. for* ^{239}Pu (%)	γ *energy (MeV)*	*Abs. int.*	*Genetic relationships*
$^{129}_{52}Te$	β^-	68.7 min		0.8		0.275*	1.7%	daughter of 4.6 h ^{129}Sb
						0.455	15%	and of 37 days ^{129m}Te
						1.08	1.5%	
						others		
$^{139}_{56}Ba$	β^-	82.9 min	6.45	6.55	5.87	0.166	23%	
						others		
$^{78}_{33}As$	β^-	91 min		0.020		0.614	R 42	daughter of 1.5 h ^{78}Ge
						0.70	R 15	
						0.83	R 8	
						1.31	R 11	
						others		
$^{78}_{32}Ge$	β^-	1.47 h		0.020		0.277	94%	parent of 91 min ^{78}As
						0.294	6%	
$^{83}_{35}Br$	β^-	2.41 h	0.87	0.51	0.084	0.530	1.4%	
$^{117m}_{49}In$	β^-, IT	1.93 h		0.011		0.158	14%	daughter 3.4 h ^{117m}Cd
						0.314	31%	
$^{92}_{38}Sr$	β^-	2.7 h		5.3		0.23	3%	parent of 3.5 h ^{92}Y
						0.44	4%	
						1.37	90%	
$^{117m}_{48}Cd$	β^-, IT	3.4 h		0.011		0.273	18%	parent of 1.9 h ^{117m}In
						0.314**	8%	
						0.345	4%	
						0.565	6%	
						0.715	4%	**with ^{117m}In
						0.880	10%	
						1.065	9%	
						1.117	4%	
						1.24*	11%	
						1.338	8%	
						1.408	8%	

TABLE XVIII (continued)

Fission product	Decay mode	Half-life	F.y. for ^{233}U (%)	F.y. for ^{235}U (%)	F.y. for ^{239}Pu (%)	γ energy (MeV)	Abs. int.	Genetic relationships
						1.433	10%	
						1.562	6%	
						1.998	15%	
						2.319	3%	
$^{92}_{39}Y$	β^-	3.53 h		5.3		0.448	2.3%	daughter of 2.7 h ^{92}Sr
						0.560	2.6%	
						0.934	14%	
						1.40	4.7%	
						others		
$^{141}_{57}La$	β^-	3.87 h	7.1	6.4	5.7	1.37	2%	parent of ^{141}Ce
		4.3 h						
$^{105}_{44}Ru$	β^-	4.44 h		0.9		0.475	20%	
						0.67	16%	
						0.726	48%	
						others		
$^{73}_{31}Ga$	β^-	4.3 h		$1.1 \cdot 10^{-4}$		0.054	9%	
						0.295	94%	
						0.74	6%	
$^{135}_{53}I$	β^-	6.68 h	5.5	6.1	5.7	0.42	7%	
						0.86	11%	
						1.04	9%	
						1.14	37%	
						1.28	34%	
						1.46	12%	
						1.72	19%	
$^{156}_{62}Sm$	β^-	9 h		0.16		1.80	11%	
$^{135}_{54}Xe$	β^-	9.14 h		6.3		0.527	80%	
$^{91}_{38}Sr$	β^-	9.67 h	5.57	5.81	2.43	0.645	15%	
						0.748	27%	

TABLE XVIII (*continued*)

Fission product	*Decay mode*	*Half-life*	*F.y. for* ^{233}U *(%)*	*F.y. for* ^{235}U *(%)*	*F.y. for* ^{239}Pu *(%)*	γ *energy (MeV)*	*Abs. int.*	*Genetic relationships*
						0.93	3%	
						1.025	30%	
						1.413	5%	
$^{93}_{39}Y$	β^-	10.3 h		6.1		0.267	6%	
						0.94	2.3%	
						1.90	1.8%	
						others		
$^{77}_{32}Ge$	β^-	11.3 h	0.011	0.0031		0.21	61%	parent of ^{77}As (38.7 h)
						0.263	45%	
						0.368	15%	
						0.417	25%	
						0.563	18%	
						0.73	14%	
						others		
$^{130}_{53}I$	β^-	12.3 h		$5 \cdot 10^{-4}$		0.419	35%	
						0.538	99%	
						0.669	100%	
						0.743	87%	
						1.15	12%	
$^{109}_{46}Pd$	β^-	13.47 h	0.044	0.030	1.40	0.088	5%	
						others		
$^{157}_{63}Eu$	β^-	15.1 h		0.0078		0.055	5%	
						0.064	27%	
						0.32	5%	
						0.37	14%	
						0.413	27%	
						0.477	5%	
						0.623	6%	
$^{97}_{40}Zr$	β^-	17.0 h		5.9	5.5	0.747	92%	radiations from ^{97m}Nb included

TABLE XVIII (continued)

Fission product	Decay mode	Half-life	F.y. for ^{233}U (%)	F.y. for ^{235}U (%)	F.y. for ^{239}Pu (%)	γ energy (MeV)	Abs. int.	Genetic relationships
$^{133}_{53}I$		20.3 h		6.9	5.2	0.53	90%	
$^{96}_{41}Nb$		23.35 h	$6.5 \cdot 10^{-3}$	$6.1 \cdot 10^{-4}$	$3.6 \cdot 10^{-3}$	0.459	28%	
						0.569	59%	
						0.778	97%	
						0.811	14%	
						0.851	22%	
						1.092	49%	
						1.200	21%	
$^{131m}_{52}Te$		30 h		0.44		0.78	60%	
						0.85	31%	
						1.127	13%	
						1.206	11%	
						others		
$^{143}_{58}Ce$		33 h		5.7	5.3	0.057	11%	
						0.293	46%	
						0.668	7%	
						0.725	8%	
						others		
$^{82}_{35}Br$		35.3 h	$1.1 \cdot 10^{-3}$	$4 \cdot 10^{-5}$		0.554	66%	
						0.619	41%	
						0.698	27%	
						0.773	83%	
						0.828	25%	
						1.044	29%	
						1.317	26%	
						1.675	16%	
$^{105}_{45}Rh$		35.88 h			3.9	0.306	5%	
						0.319	19%	
$^{77}_{33}As$		38.7 h	0.021	0.0083		0.239	2.5%	also from ^{77}Ge (11.3 h)
						others		

TABLE XVIII (*continued*)

Fission product	*Decay mode*	*Half-life*	*F.y. for* ^{233}U *(%)*	*F.y. for* ^{235}U *(%)*	*F.y. for* ^{239}Pu *(%)*	γ *energy (MeV)*	*Abs. int.*	*Genetic relationships*
$^{72}_{30}Zn$	β^-	46.5 h		$1.6 \cdot 10^{-5}$	$1.2 \cdot 10^{-4}$	0.145	90%	parent of $^{72}_{31}Ge$ (14.1 h)
						0.192	10%	
						others		
$^{153}_{62}Sm$	β^-	46.8 h	0.11	0.15	0.37	0.070	5.4%	
						0.103	28%	
						others		
$^{149}_{61}Pm$	β^-	53.1 h			1.4	0.286	2%	
						others		
$^{115}_{48}Cd$	β^-	53.5 h	0.020	0.0097	0.0038	0.262	2%	
						0.49	10%	
						0.53	26%	
						others		
$^{99}_{42}Mo$	β^-	66.7 h	4.80	6.06	6.10	0.181	7%	
						0.372	1%	
						0.740	12%	
						0.780	4%	
						others		
$^{132}_{52}Te$	β^-	77.7 h	4.4	−4.7	5.1	0.053	1%	
						0.230	90%	
$^{166}_{66}Dy$	β^-	81.5 h			$6.8 \cdot 10^{-5}$	0.082	12%	
						others		
$^{127}_{51}Sb$	β^-	93 h	0.060	0.13	0.39	0.25	*R*.26	
						0.31	*R*.11	
						0.46	*R*.100	
						0.77	*R*.45	
						others		

DECAY CONSTANTS AND FISSION YIELDS (F.y.) OF THE MAIN INTERMEDIATE-LIVED FISSION PRODUCTS

Fission product	*Decay mode*	*Half-life*	*F.y. for ^{233}U (%)*	*F.y. for ^{235}U (%)*	*F.y. for ^{239}Pu (%)*	*γ energy (MeV)*	*Abs. int. (%)*	*Genetic relationship*
$^{133}_{54}Xe$	β^-	5.27 days		6.62	6.31	0.081	37	daughter of 53.8 h ^{133m}Xe
$^{161}_{65}Tb$	β^-	6.9 days		$7 \cdot 10^{-5}$	0.0039	0.057	5	daughter of 3.7 min ^{151}Gd
						0.075	10	
$^{111}_{47}Ag$	β^-	7.5 days	0.024	0.019	0.23	0.247	1	daughter of 22 min ^{111}Pd
						0.342	6	
$^{131}_{53}I$	β^-	8.05 days	2.9	~3.1	3.74	0.080	2.6	0.8% decay to 12 days ^{131m}Xe
						0.284	5.4	
						0.364	82	
						0.637	6.8	
						0.723	1.6	
$^{125}_{50}Sn$	β^-	9.4 days	0.052	0.013	0.071	0.811	1.5	parent of 2.0 yr ^{125}Sb
						0.904	1.4	
						1.068	4	
						others		
$^{147}_{60}Nd$	β^-	11.06 days		~2.7	2.2	0.091	28	daughter of 12 min ^{147}Pr
						0.319	3	
						0.43*	4	
						0.533	13	
$^{140}_{56}Ba$	β^-	12.8 days	5.4	6.35	5.4	0.163	6	daughter of 66 sec ^{140}Cs; parent of 40 h ^{140}La
						0.305	6	
						0.438	5	
						0.537	34	
$^{136}_{55}Cs$	β^-	13.7 days	0.12	0.0068	0.11	0.067	11	radiations from daughter 0.32 sec ^{136m}Ba are included in this list
						0.086	6	
						0.16*	36	
						0.273	18	
						0.340	53	
						0.818	100	
						1.05	82	
						1.25	20	

TABLE XIX (*continued*)

Fission product	*Decay mode*	*Half-life*	*F.y. for* ^{233}U *(%)*	*F.y. for* ^{235}U *(%)*	*F.y. for* ^{239}Pu *(%)*	*γ energy (MeV)*	*Abs. int. (%)*	*Genetic relationship*
$^{156}_{63}Eu$	β^-	15.4 days	0.011	0.014	0.11	0.089	8	daughter of 9 h ^{156}Sm
						0.646	7	
						0.723	6	
						0.812	9	
						1.07*	11	
						1.15*	14	
						1.24*	16	
						1.97*	7	
						2.098	3	
						2.19*	5	
$^{86}_{37}Rb$	β^-	18.66 days	$2.3 \cdot 10^{-4}$	$2.9 \cdot 10^{-5}$	$2.3 \cdot 10^{-5}$	1.078	8.8	
$^{141}_{58}Ce$	β^-	32.5 days		~6	5.1	0.145	48	daughter of 3.8 h ^{141}La
$^{129m}_{52}Te$	IT, β^-	34.1 days		0.35		0.69	6	parent of 72 min ^{129}Te; daughter of 46 h ^{129}Sb
$^{95}_{41}Nb$	β^-	35 days	6.1	6.2	5.8	0.765	100	daughter of 66 days ^{95}Zr
$^{103}_{44}Ru$	β^-	39.5 days	1.8	3.0	5.67	0.497	88	both are daughter radiations from 5.7 min ^{103m}Rh
						0.610	6	
$^{115m}_{48}Cd$	β^-	43 days	0.0011	0.0031	0.0001	0.935	1.9	
						others		
$^{95}_{40}Zr$	β^-	65.5 days	6.1	6.2	5.8	0.724	49	parent of 35 days ^{85}Nb
						0.756	49	
$^{144}_{58}Ce$	β^-	284 days	4.5	~6.0	3.79	0.080	2	parent of 17 min ^{144}Pr; radiations of ^{144}Pr have been included in this list
						0.134	11	
						0.695	1.5	

* Complex.

TABLE XX

DECAY CONSTANTS AND FISSION YIELDS (F.y.) OF THE MAIN LONG-LIVED FISSION PRODUCTS

Fission product	*Decay mode*	*Half-life*	*F.y. for ^{233}U (%)*	*F.y. for ^{235}U (%)*	*F.y. for ^{239}Pu (%)*	*γ energy (MeV)*	*Abs. int. (%)*	*Genetic relationships*
$^{106}_{44}Ru$	β^-	368 days	0.24	0.38	4.57	0.512	21	all the listed radiations are from its 30 sec ^{106}Rh daughter
						0.622*	11	
						1.05*	1.5	
						others		
$^{155}_{63}Eu$	β^-	1.8 yr		0.033		0.087	32	
						0.105	20	
$^{125}_{51}Sb$	β^-	2.71 yr		0.021		0.176	6	daughter of 9.6 days ^{125}Sn; 30% decay to 58 days ^{125m}Te
						0.427	31	
						0.463	10	
						0.599*	24	
						0.634	11	
						0.66*	3	
$^{137}_{55}Cs$	β^-	30 yr	6.58	6.15	6.63	0.662	85	the listed radiation is from 2.5 min ^{137m}Ba daughter

* Complex.

a gamma emitting nuclide with a half-life shorter than 1 h is supported by a longer lived parent, its radiations have been included in the list of the parent and a note has been added in the genetic relationships column. Only gamma radiations having an absolute intensity higher than 1% have been reported; when many other gammas are present, they are marked "others" at the bottom of the list. Absolute intensities are usually reported; when only relative intensities are available they have been reported preceded by *R*. Half-lives, gamma energies and absolute (or relative) intensities are from LEDERER et al. (1967); fission yields from HYDE (1964).

In addition to the reported nuclides, some other isotopes having mass number too low to be generated in appreciable amount as direct fission product, are found among the fall-out products. They are believed to originate as neutron activation products after the nuclear explosion. Table XXI reports the main radioactive

TABLE XXI

DECAY CONSTANTS OF THE MAIN LOW Z RADIONUCLIDES OBSERVED IN FALL-OUT

Isotope	*Half-life*	*Decay mode*	*Energy (MeV)*	*Absolute intensity (%)*
$^{58}_{27}Co$	71.3 days	EC, β^+	0.511	30
			0.810	99
			0.865	1.4
			others	
$^{88}_{39}Y$	108.1 days	EC, β^+	0.898	91
			1.836	100
$^{57}_{27}Co$	270 days	EC	0.122	87
			0.136	11
			others	
$^{54}_{25}Mn$	303 days	EC	0.835	100
$^{134}_{55}Cs$	2.046 yr	β^-	0.57*	20
			0.605	98
			0.796*	99
			1.038	1
			1.168	1.9
			1.365	3.4
$^{60}_{27}Co$	5.263 yr	β^-	1.173	100
			1.332	100

* Complex.

characteristics of the principal low mass number gamma emitters observed in fall-out.

The production and atmospheric transport of radioactive fall-out has been recently reviewed in a Conference held at Germantown, Maryland on November 3–6, 1964 (*Radioactive Fall-out from Nuclear Weapons Tests, Proceedings of the Conference*, U.S.A.E.C., 1965). Intermediate and long lived products, undergoing a world-wide transportation through the atmosphere, are the main causes of interference with the gamma spectrometry of natural radio-activity in the field. As pointed out in Chapter 3, they can constitute a serious disturbance in total radiometric surveys. Lateral concentrations by rainwaters of artificial nuclides in surface soils can simulate natural anomalies, as observed in some cases (GASPARINI et al., 1963; ADAMS, 1966; GASPARINI and LUONGO, 1965). CERRAI et al. (1964) have reported the presence of gamma emitting fall-out products in the uppermost sediments of the Ligurian Sea. Typical fission products gamma spectra obtained in soils are reported in Appendix 2.

CHAPTER 5

Calibration of Laboratory Spectrometers and Analysis of Spectra

GAMMA SPECTRA

The shape of the spectrum of a monoenergetic gamma-ray beam is dependent on the photon energy, on the absorbing characteristic of the detector, and on the decay mode of the emitter. Furthermore, the presence of any material around the source or the detector also has an important influence on the spectra as discussed in Chapter 3. However, where different monoenergetic point emitters are measured with the same detecting assembly, the spectrum shape varies only as a function of the photon energy, count rate, and decay mode of the emitter.

With α and soft β^- particles, no noticeable interference from primary particles occurs and the spectrum shape is governed solely by the relative probabilities of the three main absorption processes. Where photoelectric absorption predominates (see Fig.7), the pulse height distribution consists essentially of the full energy peak and of the X-ray escape peak. The full energy peak is almost exclusively due to photoelectric events and the relative height of the two peaks is dependent on the photon energy and on the detector. The spectrum is essentially that reported in Fig.69 after CROUTHAMEL (1960), showing a NaI(Tl) spectrum of ^{239}U, a nuclide with only one gamma emission of 0.074 MeV. Where the Compton events predominate, the X-ray escape is negligible, at least for solid detectors (see Fig.25); the spectrum consists of the full energy peak and of a Compton continuum ranging from zero up to the full energy peak. Fig.70 (from HEATH, 1957) is a representative spectrum of a monoenergetic gamma emitter of intermediate energy (^{54}Mn). The full energy peak is formed by single photoelectric as well as by multiple Compton and photoelectric events and the Compton continuum is formed by photons scattered out of the detector after one or more Compton events and by photons whose energy has been degraded by scattering from the material surrounding the detector. The Compton continuum has two broad maxima: (*1*) at the Compton edge, whose energy can be evaluated from the initial energy of the photons with eq.(1.14); and (*2*) below 0.25 MeV, due to photons that have been back-scattered 180° from the material surrounding the detector. The energy of the backscattering peak can be evaluated with the plot of Fig.9 from the initial energy of the photons. Above 1.02 MeV pair production absorption begins and additional gaussian peaks occur at 0.511 and 1.02 MeV, as measured down from the full energy peak. The full peak minus 0.511 MeV peak occurs where

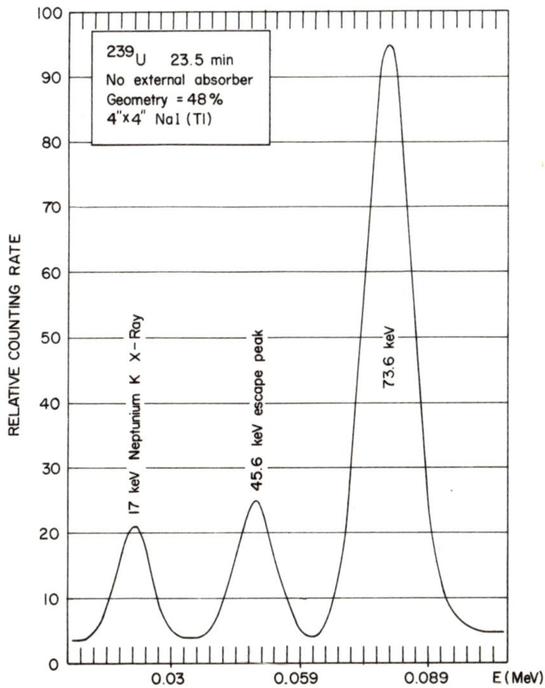

Fig.69. Gamma scintillation spectrum of ^{239}U. (After CROUTHAMEL, 1960; courtesy of Pergamon Press and of the author.)

one of two annihilation photons is not detected; the full energy peak minus 1.02 MeV occurs where the pair is not detected. A typical case is shown in Fig.71 (from HEATH, 1957), reporting the NaI(Tl) spectrum of ^{28}Al; escape peaks occur at 1.27 and 0.76 MeV. Pair production escape peaks occur in the ^{232}Th series spectrum (Appendix 2) due to the 2.615 MeV from ^{208}Tl. Some excited nuclides emit two (or more) gamma rays in cascade (e.g., ^{60}Co and ^{208}Tl). In this case, if the detector fully absorbs both gamma rays, they are not resolved and sum up to give a "sum peak". This is the case with the 1.17 and 1.33 MeV gamma rays from ^{60}Co (sum peak = 2.50 MeV) and 2.615 and 0.583 MeV from ^{208}Tl (sum peak = 3.20 MeV, see Fig.48). The full energy sum peaks are accompanied by a lower energy continuum due to summing up of Compton scattered gamma rays.

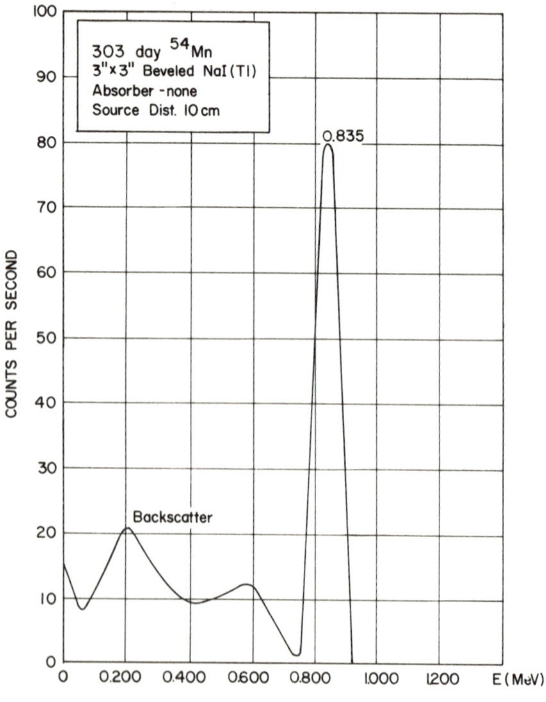

Fig.70. Gamma scintillation spectrum of ^{54}Mn. (After HEATH, 1957.)

Where the gamma emitter decays by hard β^- emission, an intense bremsstrahlung X-ray spectrum may dominate the low energy part of the spectrum. The typical shape of bremsstrahlung spectra is shown in Fig.74, which is a NaI(Tl) spectrum of a source of ^{204}Tl, a nuclide emitting β^- rays with 0.76 MeV maximum energy and no gammas. The spectrum ranges from zero up to the maximum β^- energy, but is practically negligible above 0.5 MeV. Positron energy is usually low and positron emission is not a cause of intense bremsstrahlung; however, the annihilation of positrons produces characteristic annihilation quanta full energy or sum peaks (see Fig. 72 relative to ^{22}Na, from HEATH, 1957).

Prominent K or L X-ray full energy peaks occur after electron capture decay or internal conversion de-excitation (see Fig.73

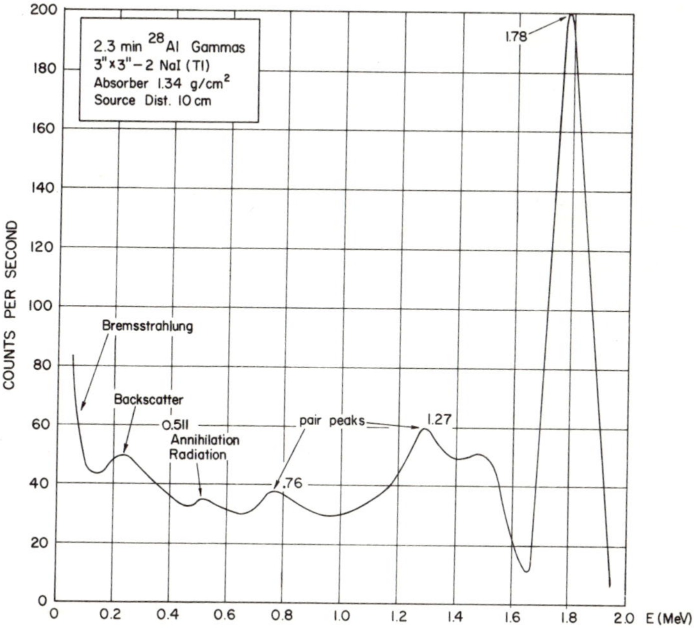

Fig.71. Gamma scintillation spectrum of ^{28}Al. (After HEATH, 1957.)

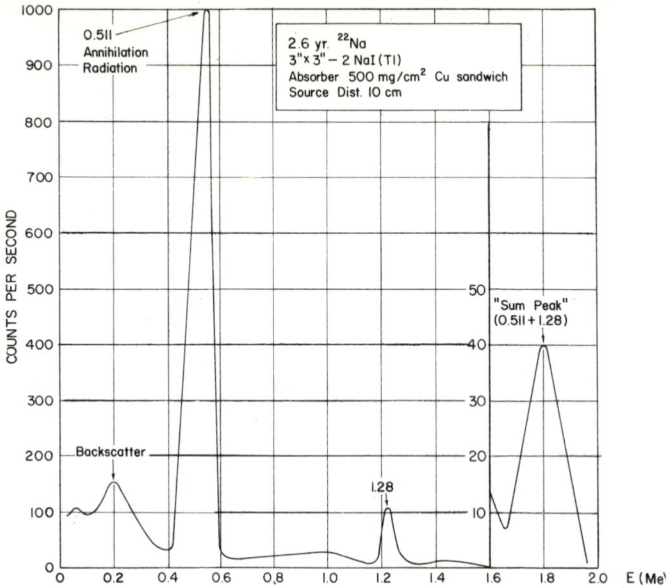

Fig. 72. Gamma scintillation spectrum of ^{22}Na. (After HEATH, 1957.)

relative to ^{137}Cs, after CROUTHAMEL, 1960, where K X-rays from ^{137}Ba are due to internal conversion of ^{137m}Ba).

Where a source emits many energetically different groups of gamma rays the spectrum is of course more complicated (see several of the spectra in Appendix 2). In any case the occurrence of clearly distinguishable gaussian peaks allows one to still identify the different gamma ray groups, which is necessary for any evaluation of the emission intensities. Both operations can be performed only after a calibration of the spectrometric equipment.

ENERGY CALIBRATION

The energy calibration consists of finding out the dependence of the pulse height (usually expressed as channel number of the multianalyzer) on the energy spent by photons in the detector. In a properly devised instrument the relationship is roughly linear, with

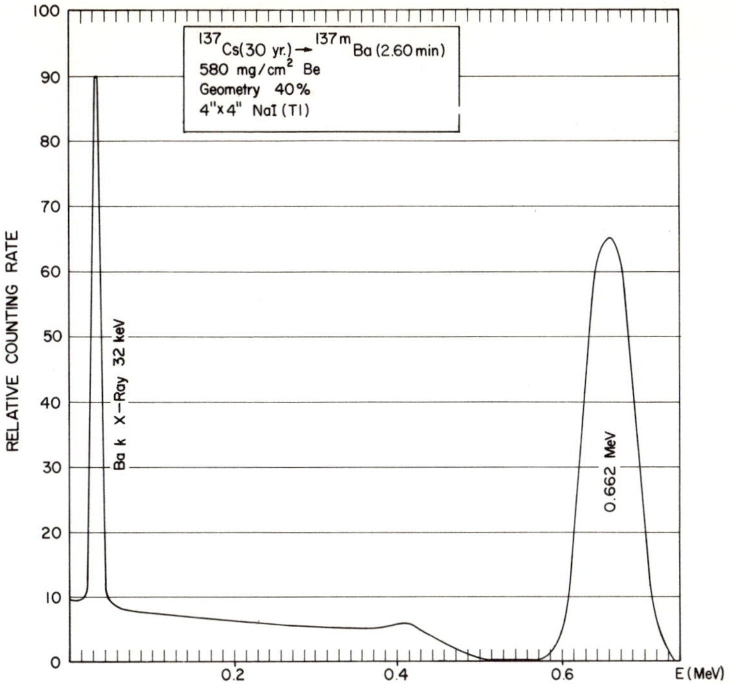

Fig.73. Gamma scintillation spectrum of ^{137}Cs. (After CROUTHAMEL, 1960; courtesy of Pergamon Press and of the author.)

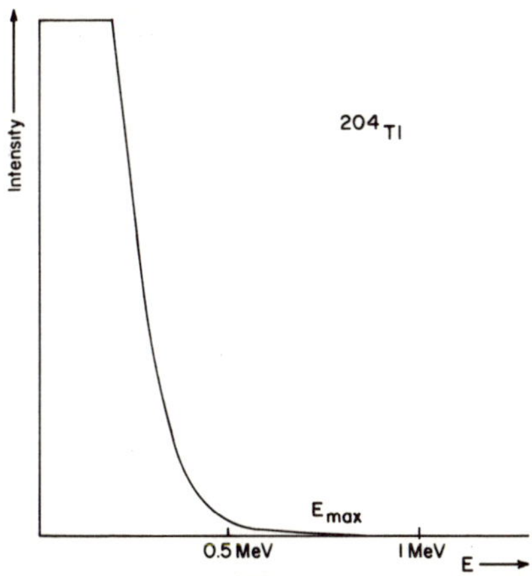

Fig.74. Bremsstrahlung spectrum of ^{204}Tl, an emitter with no gammas. (After DUIVENSTIJN and VENVERLOO, 1963; courtesy of Philips.)

a trend to increase the slope with increasing energy and major deviations below 0.1 MeV. The energy calibration is usually performed by analyzing sources emitting photons of known energy and plotting the channel number where the full energy peaks (or other known features of the spectrum) occur against the energy spent by the photons in the detector. A variety of sources suitable for these purposes are commercially available, and Table XXII lists those most commonly used. A major requirement of energy calibration standards is that the gamma energies must be known with great accuracy, usually to better than 1%. It should be noted that the pulse height vs. energy relationship is only dependent on the internal characteristics of the spectrometer and not on the geometry conditions.

A careful energy calibration is required every time an unknown spectrum is analyzed. The choice of the proper calibration standards is of course dictated from the energy range of the photons from the sample and on the precision sought. Gamma spectra of natural systems have nearly always the same characteristic shape with the same more or less pronounced full energy peaks. The energy of some of them (e.g., 2.615 MeV due to ^{208}Tl and particularly 1.461 MeV due to ^{40}K) is well enough established to allow their use as internal energy calibration standards.

The pulse height vs. energy curve is not constant with the time, but exhibits a more or less pronounced instrumental drift. This drift can consist either of a parallel displacement (zero intercept drift) and/or a variation of the slope (slope drift). These can be due to inadequate stabilization of the electronic components and to temperature changes. Where long counts are carried out, measurable gain drifts have the effects of broadening and lowering all full energy peaks, and smoothing out the Compton continuum. These effects can cause serious errors in the background subtraction and the analysis of spectra, particularly where simple computer techniques using a preset number of channels are applied. A combination of high stability electronic components with a temperature stabilized room can reduce the gain drifts to levels of about 1% in 200 h or less. It is, however, advisable to check before and after each run the position of known full energy peaks. Because of the possibility of a slope drift, at least two points should be checked.

TABLE XXII

SOME ENERGY CALIBRATION STANDARDS
(Data from LEDERER et al., 1967)

Isotope	*Half-life*	*Gamma-ray energy (keV)*
^{109}Cd	458 days	25.0 (±0.1)
		87.7 (±0.2)
^{57}Co	270 days	121.97 (±0.05)
		136.33 (±0.04)
^{207}Bi	30 yr	569.63 (±0.08)
		1063.58 (±0.06)
		1769.71 (±0.13)
^{110m}Ag	253 days	657.61 (±0.15)
		677.36 (±0.20)
		686.80 (±0.25)
		706.28 (±0.25)
		743.99 (±0.25)
		763.77 (±0.22)
		817.87 (±0.30)
		884.46 (±0.25)
		937.2 (±0.3)
		1383.8 (±0.4)
		1475.5 (±0.4)
		1504.6 (±0.5)
		1561.8 (±0.4)
^{137}Cs	30.0 yr	661.595 (±0.076)
^{54}Mn	303 days	834.85 (±0.10)
^{65}Zn	245 days	1115.44 (±0.01)
^{60}Co	5.26 yr	1173.226 (±0.040)
		1332.483 (±0.046)
^{22}Na	2.60 yr	511.006 (±0.02)
		1274.53 (±0.10)
^{40}K	$1.30 \cdot 10^9$ yr	1460.75 (±0.06)*
^{232}Th source	$1.41 \cdot 10^{10}$ yr	238.61 (±0.01) from ^{212}Pb
		583.139 (±0.023) from ^{208}Tl
		727.1 (±0.01) from ^{212}Bi
		911.07 (±0.07) from ^{228}Ac
		2614.47 (±0.10) from ^{208}Tl
^{226}Ra source	1602 yr	241.924 (±0.03) from ^{214}Pb
		295.217 (±0.039) from ^{214}Pb
		351.992 (±0.062) from ^{214}Pb
		609.37 (±0.16) from ^{214}Bi
		1120.42 (±0.46) from ^{214}Bi
		1764.45 (±0.22) from ^{214}Bi
		2204.25 (±0.48) from ^{214}Bi

* After KING et al. (1967).

For quick energy calibration purposes, it can be useful to have in the laboratory at least several high activity emitters of different energies and long enough half-lives to be usable for some years. ^{137}Cs, ^{60}Co, ^{40}K, and supported ^{208}Tl sources are quite adequate for these purposes.

ABSOLUTE CALIBRATION AND SPECTRAL ANALYSIS

General case

Let us consider a point monoenergetic gamma-ray source emitting P photons per unit time. The number of counts, n, per unit time over the whole spectrum will be:

$$n = \omega \varepsilon_i P \tag{5.1}$$

where ε_i is the detector's intrinsic efficiency and ω the geometry. The number of pulses counted per unit time under the full energy peak (n_F) will be:

$$n_F = \omega \varepsilon_i P p \tag{5.2}$$

where p is the photofraction.

In dealing with complex spectra, it is usually much more convenient to evaluate n_F than n; knowing n_F, where ω, ε_i, and p are known, it is possible to estimate the emission rate. If the absolute intensities of the gamma radiations are known, the concentration of the emitter in the sample can be evaluated.

Both $\varepsilon_i p$ and ω can be determined by both theoretical and experimental methods. The theoretical approach consists of determining ω by mere geometrical considerations, while $\varepsilon_i p$ can be calculated by the Monte Carlo method. The Monte Carlo method is a computer simulation of all the events a photon can undergo upon entering a given detector. The simulations are repeated for a large number of photons of a given energy (BERGER and DOGGETT, 1956; MILLER and SNOW, 1960b).

A more direct method is to determine experimentally all three parameters at once. This can be done by measuring standards and samples under exactly the same geometric conditions with the same detector assembly. Table XXIII reports some commercially avail-

TABLE XXIII

SOME STANDARDS FOR ABSOLUTE CALIBRATION AND SHAPE

Isotope	*Half-life*	*Gamma-ray energy*	*Absolute intensity (%)*	*Use*
^{139}Ce	140 days	0.165		shape
^{203}Hg	46.9 days	0.279	77	shape, abs.
^{22}Na	2.60 yr	0.511	178	shape, abs.
		1.275	100	
^{137}Cs	30.0 yr	0.662	85	shape, abs.
^{54}Mn	303 days	0.835	100	shape, abs.
^{88}Y	103 days	0.898	91	shape, abs.
		1.836	100	
^{65}Zn	244 days	1.112		shape
^{60}Co	5.26 yr	1.174	100	abs.
		1.332	100	
^{40}K	$1.30 \cdot 10^9$ yr	1.461	11	shape, abs.

able standards that can be used for this purpose. The usual way of performing an absolute calibration is to determine n_F/P for a number of different gamma-ray energies and to plot the obtained values in a full energy absolute efficiency ($\omega\varepsilon_i p$) vs. gamma-ray energy plot. Although amplitude comparison of full energy peaks is permissible in principle where gamma-ray energies are very close, it is advisable to compare always the areas of full energy peaks. This in fact improves the counting statistics considerably, because a larger part of the spectrum is considered, and reduces errors due to lowering and broadening of the Gaussian peaks by gain drifts.

Where the sample emits gamma rays of different energies, all the full energy peaks except the highest energy one, are superimposed on the Compton continuum due to higher energy gamma rays. In other words, all the full energy peaks but one are formed not only by the full energy events, but also by higher energy photons that have not been totally absorbed by the detector. Furthermore, bremsstrahlung X-rays are important below 0.5 MeV where β^- emission also occurs. It is therefore necessary to correct these full energy peaks for the Compton and bremsstrahlung contributions. With high resolution detectors, such as Ge(Li) semiconductors, the energy interval covered by a full energy peak is usually sufficiently small so as to permit the graphical subtraction of the Compton and bremsstrahlung contributions. Where full energy peaks are not close to each other, a least mean square fit can be applied to the continuum and the graphical subtraction performed with high accuracy. Higher errors have to be expected when the full energy peak is located close to some Compton edge. This procedure is not always possible with scintillation detectors or where a series of

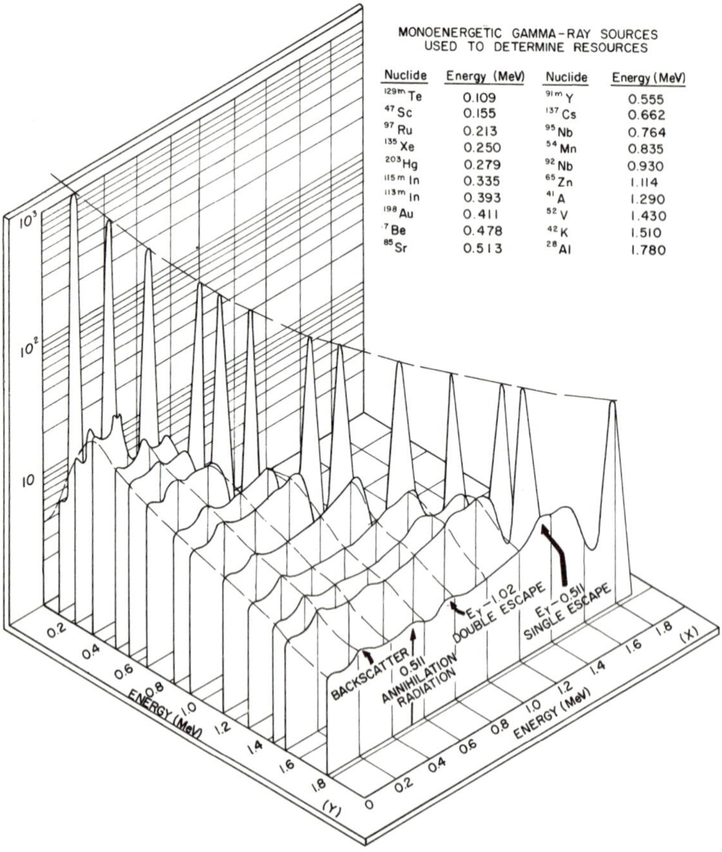

Fig.75. Three dimensional representation of the variation of the gamma spectrum of monoenergetic emitters with the photon's energy. The horizontal axes represent the energy of the gamma emission (Y) and the energy distribution of individual spectra as recorded at the multichannel analyzer. The vertical axis represents the counting rate. (After HEATH, 1966; courtesy of North-Holland Publ. Co.)

closely spaced full energy peaks completely masks the continuum. In such cases the spectrum shape for each gamma energy must be inferred by calibration and a step by step calibration procedure must be consequently applied. The calibration can be performed by measuring a number of monoenergetic gamma-ray shape standards of different energies. The evolution of the spectrum shape with the photon energy of monoenergetic emitters with no major contributions from primary decay modes is like that reproduced in Fig.75 (from HEATH, 1961, 1966). It is a three dimensional model of a 3 × 3 inch (7.6 × 7.6 cm) NaI(Tl) spectrometer response for gamma rays ranging from 0.109 to 1.780 MeV. The sources are listed in the figure. Additional sources that can be used as shape standard are listed in Table XXIII and designated "shape." It is, of course, unnecessary that the shape standards be absolutely calibrated, but, in order to guarantee directly comparable results, they must have comparable emission intensities and be analyzed with exactly the same detecting equipment and geometry. Where a suite of monoenergetic spectra covering the whole energy range of interest is measured, the ratio of the full energy peak area to the spectrum height (R) can be evaluated for discrete pulse height values and plotted against the respective gamma-ray emission energies. A set of curves is so obtained, each of them representing the variation of R for a given pulse height against the gamma-ray energy. The complex spectrum is then unscrambled step by step starting from the highest energy full energy peak, and rebuilding the spectrum shape relative to each full energy peak upon the above described diagram. VAN LIESHOUT et al. (1965) describe an analogous graphical peeling method consisting of graphical subtraction of standard spectra after normalization of the intensities.

When the bremsstrahlung X-ray spectrum must also be accounted for, a further shape calibration is necessary. Bremsstrahlung shapes can be evaluated by analyzing pure β^- emitters of different energies and reconstituting the spectra, eventually by interpolation, upon the obtained library of spectra.

Ordinary rock analysis

All the above described calibration procedures are based upon the assumption that the interaction of gamma rays with the source itself are negligible. Where rock samples are analyzed, they are usually crushed and placed around the detector in a thick (several cm) uniform layer. Solid rock samples are sometimes used. The greatest advantage of gamma over alpha-ray spectrometry lies in the fact that thick sources can be used with confidence. As a consequence, no radiochemical treatment of the sample is required, because the counting rate can be increased by using larger samples rather than by preconcentrating the radioelements. The presence of a thick source close to the detector produces, however, some secondary effects that need to be considered if an accurate determination is sought. The main effects are: (*a*) variation of the background; (*b*) deformation of the spectrum; (*c*) self absorption.

A thickness of a few cm of ordinary rock sample has of course no significant effect on the penetrating cosmic ray component of the background, but it absorbs significantly gamma rays from environmental impurities. As a consequence a lowering of all the full energy peaks due to environmental radioactivity has to be expected. Where a background spectrum is obtained without any absorber around the detector, and this background is subtracted from a thick source spectrum, a lower than true source counting rate will be consequently obtained for the corresponding energy ranges. Furthermore, a thick source around the detector enhances the possibility that gamma rays are at least once Compton scattered before reaching the detector. This causes a lowering of all the full energy peaks and an increase of the Compton continuum height, i.e., a decrease of the photofraction. Such effects are, of course, more prominent for low energy gamma rays.

The self absorption by the source is most important where the low energy part of the spectrum is considered, or when dealing with very high density material. WATT and RAMSDEN (1964) have derived an equation for the evaluation of the transparency of a thick source to its own radiations. If t is the thickness of the source and μ its total absorption coefficient, the fraction F of gamma rays transmitted is given by:

$$F = \frac{1 - \exp(-\mu t)}{\mu t} \tag{5.3}$$

Fig.76 reports F as a function of $\mu_m t$ (where μ_m is the total absorption coefficient expressed in cm^2/g and t is consequently expressed in g/cm^2). Small variations of thickness can be seen to produce no remarkable changes of F when $\mu_m t$ is smaller than 0.5 or bigger than 10.

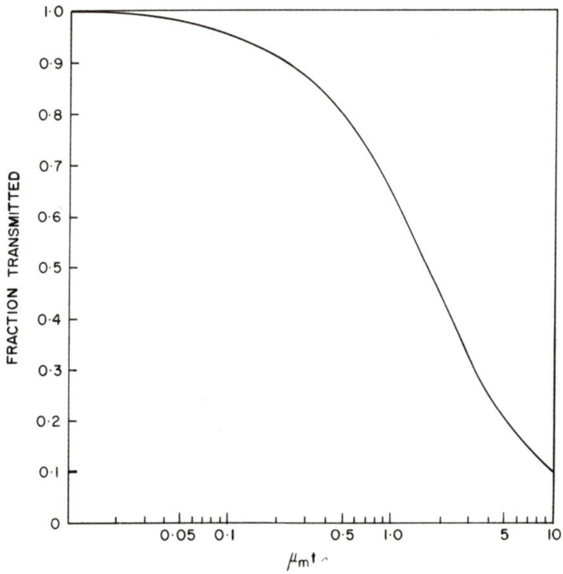

Fig.76. Source self absorption. Variation of the fraction of transmitted gamma rays as a function of $\mu_m t$. (After WATT and RAMSDEN, 1964; courtesy of Pergamon Press.)

Thick source effects can be conveniently accounted for if calibration standards having the same absorbing characteristics of the sample are used and if the background is determined surrounding the detector with a proper unactive material. A mock up having shape, atomic number and mass number very close to that of the sample can be made of suitably alternated high density PVC slabs and stainless steel plates (care should be taken that the latter be free of ^{60}Co impurities) as used by TOBAILEM et al. (1967) for the determination of the activity due to long lived nuclides in the St. Severin meteorite.

When dealing with rock analysis, it is more practical to overcome thick source calibration difficulties by diluting a high activity standard with a very low activity rock, such as dunite or other ultramafic rocks or some types of calcareous rocks, and to analyze them in canisters identical to those used for the sample. When the low energy part of the spectrum is considered, the closeness of reproducibility of absorbing characteristics between standards and samples becomes very critical. In this case the use of identical canisters is not a sufficient guarantee, an important role being played by small differences in degree of compaction and densities. Different concentrations of high $Z(>70)$ elements in the standards and in the sample have a twofold influence on the low energy spectrum. High Z elements decrease the transmittivity of low energy photons by an increased photoelectric absorption and, on the other hand, are a source of fluorescence K X-rays (see CHARLET, 1968). When dealing with energies lower than 0.5 MeV the use of solid samples may become more convenient, since it is easier to obtain reproducibility. Even if they are irregularly shaped, a mock up can be made and standard sources with very close characteristics can be prepared by using a proper cementing material. Where the volume of the sample is small relative to the volume of the NaI(Tl) crystal, irregularities in sample shape are unimportant (CHERRY and ADAMS, 1963). Standards and samples must have a comparable gamma emission rate. With high emission rates there is an increased probability that photons, emitted from different nuclides, can sum up and give rise to coincident higher energy sum peaks.

A typical NaI(Tl) scintillation and Ge(Li) spectra of rock samples are shown in Appendix 2. They are essentially formed by the superposition of the spectra of ^{232}Th and ^{238}U series and of ^{40}K.

The aim of most gamma spectrometry of rocks is the determination of the ^{238}U, ^{232}Th, and ^{40}K contents. However, only ^{40}K is a direct gamma emitter and therefore can be determined directly; ^{238}U and ^{232}Th determinations are possible only with the knowledge of the degree of equilibrium with the respective gamma emitting daughters.

Table XXIV reports the nuclides contributing to different energy ranges and shows that the simplest approach is to consider the 0.97–2.62 MeV energy range and to determine the total amounts of U, Th and K in the sample by the equations:

TABLE XXIV

MAIN CONTRIBUTORS TO THE GAMMA-RAY SPECTRUM OF A COMMON ROCK

Energy Range (MeV)	*Contributors*	*Series*	*Position in the series*	*Remarks*
2.615–2.38	^{208}Tl	^{232}Th	^{220}Rn daughter	
2.38–1.46	^{208}Tl	^{232}Th	^{220}Rn daughter	
	^{214}Bi	^{238}U	^{222}Rn daughter	
1.46–0.97	^{208}Tl	^{232}Th	^{220}Rn daughter	
	^{214}Bi	^{238}U	^{222}Rn daughter	
	^{40}K			
0.97–0.4	^{208}Tl	^{232}Th	^{220}Rn daughter	
	^{214}Bi	^{238}U	^{222}Rn daughter	
	^{40}K			
	^{228}Ac	^{232}Th	pre-^{220}Rn member	
0.4–0.09	^{208}Tl	^{232}Th	^{220}Rn daughter	
	^{214}Bi	^{238}U	^{222}Rn daughter	
	^{40}K			
	^{228}Ac	^{232}Th	pre-^{220}Rn member	
	^{212}Pb	^{232}Th	^{220}Rn daughter	from 0.25 MeV down
	^{214}Pb	^{238}U	^{222}Rn daughter	from 0.35 MeV down
	^{226}Ra	^{238}U	^{222}Rn parent	from 0.19 MeV down
	bremsstrahl.	^{232}Th ^{238}U, ^{40}K,	from all hard β^- emitters	
	^{235}U, ^{231}Pa	^{235}U		
0.09	^{208}Tl	^{232}Th	^{220}Rn daughter	
	^{214}Bi	^{238}U	^{222}Rn daughter	
	^{40}K			
	^{228}Ac	^{232}Th	pre-^{220}Rn member	
	^{212}Pb	^{232}Th	^{220}Rn daughter	
	^{214}Pb	^{238}U	^{222}Rn daughter	
	^{226}Ra	^{238}U	^{222}Rn parent	
	bremsstrahl.	^{232}Th ^{238}U, ^{40}K	from all hard β^- emitters	
	^{228}Th	^{232}Th	pre-^{220}Rn member	
	^{234}Th	^{238}U	parent of ^{234}U	
	K X-rays	^{238}U, ^{232}Th	mostly from ^{220}Rn and ^{222}Rn daughters	
	^{235}U, ^{231}Pa	^{235}U		

$$\text{Th} = (1/D_{\text{Th}})(A_1 - bk_1) = (1/D_{\text{Th}})A_{\text{Th}} \quad (5.4)$$

$$\text{U} = (1/D_{\text{U}})(A_2 - S'_{\text{Th}}A_{\text{Th}} - bk_2) = (1/D_{\text{U}})A_{\text{U}} \quad (5.5)$$

$$\text{K} = (1/D_{\text{K}})(A_3 - S''_{\text{Th}}A_{\text{Th}} - S''_{\text{U}}A_{\text{U}} - bk_3) = (1/D_{\text{K}})A_{\text{K}} \quad (5.6)$$

where:

A_1 = total counting rate in the energy range 2.38–2.62 MeV (Th range);

A_2 = total counting rate from the 2.2 MeV full energy peak to the 1.46 MeV full energy peak (U range);

A_3 = total counting rate from 1.46 MeV full energy peak to 0.97 MeV full energy peak (K range);

D_{Th} = efficiency factor for the ^{232}Th series (counting rate for unit weight of ^{232}Th) over the Th range;

D_{U} = efficiency factor for the ^{238}U series over the U range;

D_{K} = efficiency factor for K over the K range;

S'_{Th} = stripping factor for the ^{232}Th series over the U range (i.e., ratio of the ^{232}Th series counting rate over the U range to A_{Th});

S''_{Th}, S''_{U} = stripping factors for the ^{232}Th and the ^{238}U series respectively over the K range;

bk_1, bk_2 and bk_3 = background counting rates over the Th, U and K ranges respectively;

A_{Th}, A_{U}, A_{K} = net ^{232}Th, ^{238}U, and K counting rates over the Th, U, and K ranges respectively.

The efficiency and stripping factors must be determined experimentally by calibration. This has been substantially the approach in the early application of gamma spectrometry to the U, Th, and K determinations in rocks (ADAMS et al., 1958) and is still the method used in most laboratories. HURLEY (1956) suggests a similar method but he uses the 0.238 MeV full energy peak for Th, the 0.163 MeV region for U, and 1.38 MeV region for K. Only small crystals were available at that time and they were not capable of determining such full energy peaks as the 2.615 MeV from ^{208}Tl in reasonable counting times. Most of the determinations of the Th, U, and K ranges are restricted to the 2.615 MeV, 1.76 MeV, and 1.46 MeV full energy peaks respectively. The extended ranges have the substantial advantage of improving the counting statistics because a larger portion of the spectrum is considered. The K range has been extended down to 0.97 MeV, because from this energy down contribution from a pre-^{220}Rn member of the ^{232}Th series (^{228}Ac) occur, and in this way the K determination is completely independent from the equilibrium conditions in both series. In fact the contributions from ^{228}Ac can be accounted for only with equilibrium assumptions and, although only a few disequilibrium cases have been reported in the ^{232}Th series as yet, the occurrence of ^{220}Rn in the atmosphere (ISRÄEL, 1964a,b; ISRÄEL and ISRÄEL, 1965) suggests that deviations from equilibrium should occur, at least in soils on a small scale. Equilibrium can be assumed with confidence in the ^{232}Th series in most unaltered rocks and the K range can be extended in most cases down to the ^{226}Ra full energy peak (0.186 MeV). The main limitation of this method is in the ^{238}U determination, because it is based upon the assumption of equilibrium with a distant member of the series (^{214}Bi) and, as discussed in Chapter 4, many cases of disequilibrium in natural systems have been reported. ^{238}U determinations performed in this way are therefore reliable only for unaltered, more than 1 megayear old rocks.

Where the assumption of secular radioactive equilibrium in the ^{238}U series is not permissible, the ^{238}U content can be determined upon the 0.09 MeV peak, where radiations from ^{234}Th contribute. Equilibrium from ^{238}U to ^{234}Th is practically established within 160 days (see Fig.58). The total ^{238}U content of the sample is given by:

$$\mathrm{U} = \frac{1}{D_{\mathrm{UT}}}\left(A_{\mathrm{UT}} - A_{\mathrm{Th}}S'''_{\mathrm{Th}} - A_{\mathrm{U}}S'''_{\mathrm{U}} - A_{\mathrm{K}}S'''_{\mathrm{K}} - bk'''\right) \qquad (5.7)$$

where: D_{UT} = efficiency factor for ^{238}U in the 0.09 MeV full energy peak; A_{UT} = total counting rate under the 0.09 MeV peak; S'''_{Th}, S'''_{U}, S'''_{K} = stripping factors of the ^{232}Th series and of the post ^{226}Ra part of the ^{238}U series and of potassium under the 0.09 MeV peak; bk''' = background counting rate under the 0.09 Mev peak.

This method was first used by AVAN and KELLER (1961a,b). CHEMINEE and NORDEMANN (1963), BONASIA and GASPARINI (1965), CIVETTA et al. (1965) and LUONGO and RAPOLLA (1965) have applied it to U determinations in Quaternary volcanic rocks. In addition to the above described difficulties, in calibration and reproducibility, the use of this low energy part of the spectrum has the drawback that the ^{234}Th contribution is only a fraction of the total (see Table XXIV). The influence of higher energy photons Compton scattered by the shield and of X-rays from the shield can be reduced by increasing the distance of any shielding from the detector. A further improvement can be obtained by the use of small size crystals, as described by DOMINGOS and MELO (1967). They have used two different detectors: a $\frac{1}{4}$ inch (0.65 cm) thick by $1\frac{3}{4}$ inch (4.55 cm) diameter NaI(Tl) for analyzing the 0.05–0.1 MeV energy range, and a 3 inch (7.6 cm) by 3 inch (7.6 cm) NaI(Tl) for energies above 0.1 MeV. The use of a smaller crystal for low energies has the advantage of improving the resolution by more uniform transmission of the light flashes and of decreasing the Compton and bremsstrahlung background without any major effect on the full energy efficiency at the lower energies considered. It must be pointed out that, when the stripping factors are evaluated by Th, Ra, and K standards, they account for all significant X-ray contributions from these isotopes, i.e., bremsstrahlung and internal conversion K emission.

Similar stripping techniques have been used by MATHEVON et al. (1967) for unscrambling the Ge(Li) spectrum of a rock sample. In that case, energies lower than 0.186 MeV have not been considered and the determinations have been made upon 0.609 and 0.352 MeV full energy peaks for ^{238}U and 0.583 and 0.338 MeV full energy peaks for ^{232}Th (see spectrum in Appendix 2).

In order to apply any of the described techniques, the efficiency and the stripping factors must be evaluated by calibration with pure K, U, and Th standards; the latter ones in equilibrium with all their gamma emitting daughters. A great advantage when the efficiency factors are determined in this way is that no knowledge of the decay schemes is required.

Any salt where potassium is contained in known percentage with the natural isotopic composition can be used for the determination of K. The potassium determination in any sample is of course based upon the assumption of a constant isotopic composition. Up to the present, although cases of ^{39}K/^{41}K ratio variation in rocks have been reported (VERBEEK and SCHREINER, 1967; SCHREINER and VERBEEK, 1965) no significant alteration of ^{40}K isotopic ratio has been observed as yet.

A stripping method such as that described for the general case can of course be applied also to the rock analysis, as described by VITTOZZI and GASPARINI (1963). The method gives satisfactory results only for energies higher than 0.4 MeV, as the difficulty of a correct evaluation of the complex bremsstrahlung spectrum of all the natural β^- emitters is a drawback for its application to lower energies.

Geochemical rock standards are available that can be used for checking the method of analysis. In fact, in 1951 the United States Geological Survey prepared and distributed two geochemical rock standards: a granite (G-1) and a Triassic diabase (W-1). Major and trace elements contents have been measured in a great number of laboratories and have been periodically summarized (AHRENS, 1951; DENNEN et al., 1951; SCHLECHT and STEVENS, 1951; STEVENS, 1960; FLEISCHER and STEVENS, 1962; FLEISCHER, 1965, 1969). The exhaustion of the supply of G-1 and the need for a series of standards covering a wider range of composition led the U.S.G.S. to prepare a new series: a granite (G-2), in substitution of G-1, a granodiorite (GSP-1), an andesite (AGV-1), a peridotite (PCC-1), a dunite (DTS-1) and a basalt (BCR-1). Resumes of data concerning the new standards have been compiled by FLANAGAN (1967 and 1969).

In the meantime, further geochemical standards have been prepared by other laboratories and organizations, including the Non-metallic Standard Committee of the Canadian Association of Applied Spectroscopy, (CAAS) which prepared a syenite rock (CAAS syenite 1) and a sulphide ore (CAAS sulphide 1), the "Centre de Recherches Petrographiques et Geochimique de Nancy", France, which prepared three granites (GR, GH, GA) and a basalt (BR), and the Geological Survey of Tanganyika, which prepared a tonalite (T 1). Data on the Canadian standards have been summarized in a NSC-CAAS report, 1961, WEBBER (1965) and SINE et al. (1969), on the French standards by ROUBAULT et al. (1966) and on the Tanganyika standard by the Geological Survey of Tanganyika report, 1963. Furthermore, the United States

National Bureau of Standards has made available a number of Ra rock standards. U, Th, and K determinations of all these standards are summarized in Appendix 3. U and Th determinations are not very consistent for many standards. Most consistent results have been obtained for W-1, which standard is considered by the authors the most reliable standard for U and Th at this time.

COMPUTER TECHNIQUES

Manual reduction of spectra is sometimes a long and tedious procedure, and the development of computer techniques offers a time saving alternative for handling data, mainly when routine determinations are carried out. Modern multichannel analyzers can be equipped with perforated paper tape or magnetic tape read-out so that data can be transmitted to an electronic computer without further handling when a proper program has been developed.

Computer techniques applicable in general cases both to the identification and to quantitative analysis of gamma scintillation or semiconductor spectra have been described by many authors, including HEATH (1961, 1966); DEAN (1964); CHOY and SCHMITT (1965); SCHONFELD et al. (1966); YOUNG and SINGHAL (1966); HELMER et al. (1967a,b); MARISCOTTI (1967) and by many papers in *Application of Computers to Nuclear and Radiochemistry* (NAS, NRC, NAS-NS 3107, 1962).

Most of the methods are based upon a least mean square elaboration of the spectra. The method assumes that the data in a given spectral range can be represented by a function $\bar{y}_i$ of a set of parameters q_k. The best fitting function will be one having the least deviations of experimental data y_i from theoretical values, i.e., one for which:

$$S^2 = \sum_i w_i\{y_i - \bar{y}_i(q_k)\}^2 = \text{minimum} \tag{5.8}$$

where w_i are the weights of experimental points that for a multichannel operation (constant counting time in each channel) are inversely proportional to the total counts in that channel.

For S^2 to be a minimum:

$$\frac{dS^2}{dq_k} = 0 \tag{5.9}$$

for all k.

This method is applicable to linear functions. If the function is not linear, it must be "linearized". A most used linearization method is that by Gauss, consisting of an expansion of the function into a truncated Taylor's series. If $q_k = (q_k)_0 + dq_k$ we have:

$$\bar{y}_i(q_k) = \bar{y}(q_k)_0 + \sum_k (\delta\bar{y}_i/\delta q_k)_0\, dq_k + \text{higher derivatives} \tag{5.10}$$

Eq.(5.10) is linear when higher derivatives are neglected. Therefore:

$$S'^2 = \sum_i w_i\{y_i - \bar{y}_i(q_k)_0 - \sum_k (\delta\bar{y}_i/\delta q_k)_0\, dq_k\}^2 \tag{5.11}$$

and for S'^2 to be a minimum:

$$dS'^2/d(dq_k) = 0$$

As many equations as q_k are available, and a matrix solution is then applied which is especially suitable for computer data processing.

It is usually assumed that a gamma spectrum can be divided into energy intervals where the counting data dependence on the channel number, $N(x)$, can be represented by the sum of a Gaussian function, $G(x)$, representing the Gaussian peaks, and of a linear function representing the background plus Compton distribution, i.e.:

$$N(x) = G(x) + lx + m \tag{5.12}$$

where l and m are parameters characterizing the background + Compton distribution. A method to search automatically for peaks in the presence of background is based on the fact that, if $N(x)$ is a continuous function, its second derivative $N''(x)$ becomes independent of the background and vanishes in every interval where there is no peak (MARISCOTTI, 1967). KOWALSKI and ISENHOUR (1968) have developed a program which is a combination of residual least mean squares and iteration method. In this case the spectrum has been considered as represented by a modified hyperbolic secant, which fits a full energy peak better than a gaussian curve, and by a quadratic with a modified hyperbolic tangent multiplier (Compton + background).

The application of computer techniques to ordinary rock spectrometry has been discussed by RYBACH and ADAMS (1966) and RYBACH et al. (1966). In this case the counting rate in the *i*th channel is given by:

$$N_i = xU_i + y\mathrm{Th}_i + zK_i + u\mathrm{Ra}_i + v_i \qquad \text{weight } w_i \qquad (5.13)$$

where x, y, z, u are the ratios of U (pre-^{226}Ra members), Th, K, Ra amounts in the sample to those in the standards, U_i, Th_i, K_i, Ra_i the counting rates of U (pre-^{226}Ra members), Th series, K and Ra and its daughters in the standards and v_i is a random error. Having as many equations as channel numbers for four unknowns, the best values of x, y, z, and u can be found by a weighted least mean square application. The standard deviations of each unknown can be also evaluated.

When applying computer techniques, an imperative requirement is that no drift must occur. When routine analyses are performed a good precaution is to check the position of several well known energy peaks in the recorded spectra before the computer processing. A further requirement is that a library of standard spectra, which have been measured with high statistical accuracy, must be available in order to determine the relative parameters.

In the case of ordinary rock analysis only U, Th, K (and eventually Ra) in secular equilibrium spectra are of course required.

DETECTION LIMITS AND COUNTING ERRORS

Gamma-ray emissions from a source are random events occurring in a large number of nuclei and can be therefore described by Poisson statistics. Therefore, the probability P_x that x gamma rays for unit time will be emitted from a given number of identical radionuclides is:

$$P_x = A^x \frac{e^{-N}}{x!} \qquad (5.14)$$

where N is the average gamma emissions for unit time.

The counting error is usually expressed as standard deviation, σ:

$$\sigma = \sqrt{n} \tag{5.15}$$

where n is the total number of counts. Sometimes the variance, σ^2, and the probable error, p:

$$p = 0.6745\sqrt{n} \tag{5.16}$$

are quoted.

When the error is expressed as standard deviation, it means that it is 68.3% probable that the correct value is within the errors; 2σ increases such probability to 95.5%, and 3σ to 99.7%. The standard deviation is a simple function of the total number of counts, and it can be improved by increasing the counting time or by using larger samples.

The total counts recorded over a given spectral range for a preset counting time is the sum of the counts from the source (S) plus the background (B). In this case the standard deviation is:

$$\sigma = \sqrt{\sigma_{S+B}^2 + \sigma_B^2} \tag{5.17}$$

In most cases, B is the sum of a true background (b) and of other interfering activities from the sources (c), i.e., Compton scattered photons etc., then:

$$\sigma_B = \sqrt{\sigma_b^2 + \sigma_c^2} \tag{5.18}$$

When a net activity of the same order as the true background has to be measured, it is helpful to determine in advance the approximate sample + background counts needed to get a given standard deviation. The diagram by LOEVINGER and BERMAN (1951) (see Fig.77) is very useful for predicting the total number of counts needed for a given standard deviation against the background counts for different total/background ratios (r). It must be stressed that, for any value of r, the improvement of the standard deviation is not a linear function of the total counts (Fig.78), but it approaches asymptotically the abscissa axis of no error. A definite limit therefore exists where a small improvement in the standard deviation is obtained only at the expense of very long counting time and is hence impractical.

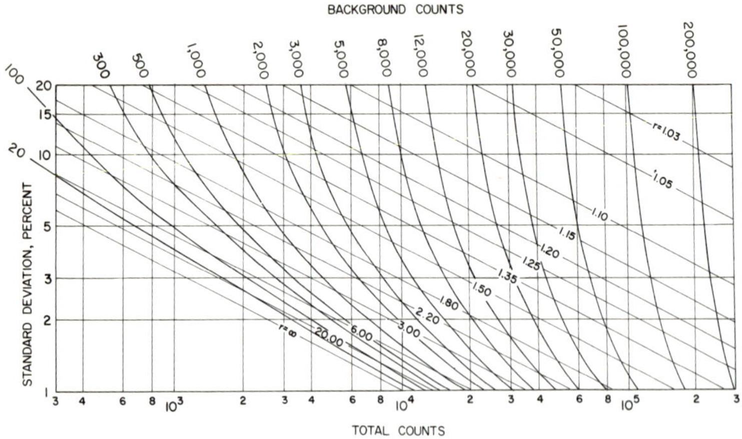

Fig.77. Diagram allowing prediction of the total number of counts needed for getting a given standard deviation, in any counting experiment, when the ratio *r* (total counting rate/background counting rate) is known. (After LOEVINGER and BERMAN, 1951.)

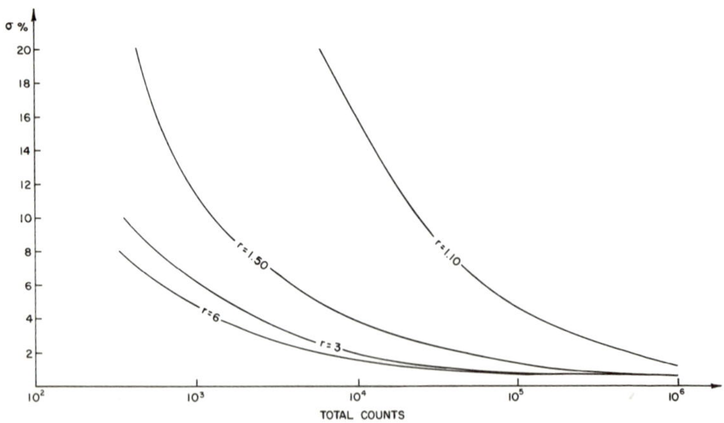

Fig.78. Variation of the standard deviation (σ) against the total number of counts for different values of r (ratio of total to background counting rate).

TABLE XXV

"WORKING EXPRESSIONS" FOR THE ESTIMATE OF THE CRITICAL LEVEL (L_C), DETECTION LEVEL (L_D), AND DETERMINATION LEVEL (L_Q) UPON THE KNOWLEDGE OF THE BACKGROUND COUNTS

(After CURRIE, 1968)

	L_C *(counts)*	L_D *(counts)*	L_Q *(counts)*
Paired observations ($\sigma_B^2 = B$)*	$2.33\sqrt{B}$	$2.71 + 4.65\sqrt{B}$	$50\left\{1 + \left(1 + \frac{B}{12.5}\right)^{\frac{1}{2}}\right\}$
"Well known" background ($\sigma_B^2 = O$)	$1.64\sqrt{B}$	$2.71 + 3.29\sqrt{B}$	$50\left\{1 + \left(1 + \frac{B}{25}\right)^{\frac{1}{2}}\right\}$
Zero background ($B = 0$)	0	2.71	100
Asymptotic ratio** (δ/σ_B)	1.64	3.29	10

* B is the "true" mean of the background.
** Well known background—correct to within 10% if $B \geqslant 0$, 6, 2,500 counts, respectively, for each of the three columns—for paired observations, $B \geqslant 0$, 36, 1,250 counts, respectively.

The problem of detection limits has been recently discussed in detail by CURRIE (1968). Three different detection levels can be defined: (*1*) a "critical level" (L_c) where detection is doubtful; (*2*) a "detection limit" (L_D) where the detection is reliable, but no quantitative estimate is possible, the counting statistics being too low; (*3*) a "determination limit" (L_Q) where quantitative analysis can be carried out.

Table XXV reports working expressions evaluated by CURRIE (1968) for the case of radioactivity counting, including neutron activation. These expressions allow the estimation of the three levels with a given background (B). In the first row (paired observations) it is assumed that two identical measurements, one for the sample plus background and the other for the background, are carried out. In the second row (well known background), it is assumed that the background has been determined with a very high statistical accuracy. The last row reports the ratio of the net counting (sample minus background) to the standard deviation of the background (σ_b) for very high number of background counts. More extensive discussions of counting errors and detection limits have been reported by WATT and RAMSDEN (1964), WING and WAHLGREN (1967), and KAISER (1965).

CHAPTER 6

Related and Collaborative Techniques

NEUTRON ACTIVATION

When a substance is bombarded with neutrons, any component nucleus may capture a neutron and is consequently transformed into a higher mass number nucleus (compound nucleus). The probability of occurrence of capture of neutrons of a given energy by a given nuclide is usually expressed as the cross section, measured in barn units. The compound nuclei are always highly excited energetically and they decrease their energy excess by prompt emission of nucleons or photons as they are transformed into a product nucleus, which in many cases has a neutron excess over stability and is therefore β^- unstable.

Neutron induced reactions are usually indicated as (n, p), (n, α), $(n, 2n)$, etc., according to the nature of the prompt emission. Where incident neutrons have energies of the order of the average kinetic energy of atoms and molecules in the sample, they are designated thermal neutrons. In this case the excess energy of the compound nucleus is not high enough to permit the emission of nucleons, and (n, γ) reactions are consequently the most probable. Many nuclei of trace elements, when bombarded with thermal neutrons, are transformed into radioactive product nuclei with short half-lives and hence high specific activity. Nuclei of elements in submicrogram concentration in any sample material can therefore be detected by means of the radioactive characteristics of product nuclei. Micro concentrations can thus be determined with good statistical accuracy after activation. In the cases of U, Th, and K the following (n, γ) reactions occur:

$$^{238}_{92}\mathrm{U} + ^{1}_{0}n \longrightarrow ^{239}_{92}\mathrm{U} \xrightarrow{\beta^-} ^{239}_{93}\mathrm{Np} \xrightarrow{\beta^-} ^{239}_{94}\mathrm{Pu} \xrightarrow{\alpha} ^{235\mathrm{m}}_{92}\mathrm{U}$$

$$^{235}_{92}U + ^{1}_{0}n \longrightarrow ^{236}_{92}U \xrightarrow{\alpha} ^{232}_{90}Th$$

$$^{234}_{92}U + ^{1}_{0}n \longrightarrow ^{235}_{92}U$$

$$^{232}_{90}Th + ^{1}_{0}n \longrightarrow ^{233}_{90}Th \xrightarrow{\beta^-} ^{233}_{91}Pa \xrightarrow{\beta^-} ^{233}_{92}U \xrightarrow{\alpha} ^{229}_{90}Th$$

$$^{41}_{19}K + ^{1}_{0}n \longrightarrow ^{42}_{19}K \xrightarrow{\beta^-} ^{42}_{20}Ca$$

Table XXVI summarizes thermal neutron activation data relative to the above reported isotopes. Extensive tables of neutron activation constants are reported by BAUMGAERTNER (1967).

^{239}Np and ^{233}Pa, because of their convenient half-lives, are the isotopes usually chosen for U and Th determination respectively. Standard procedures include the activation of samples and standards, fusion in presence of carriers, radiochemical separation, and activity measurement of ^{239}Np and ^{233}Pa. No stable isotopes of Pa and Np exist which can be used as carriers and either non-isotopic (e.g., La for ^{239}Np) or long lived isotopic carriers (e.g., ^{231}Pa and ^{237}Np) are commonly used. Details about such techniques have been reported by many authors, including MAHLMAN and LEDDICOTTE (1955); REED and TURKEVICH (1955); HAMAGUCHI et al. (1957); REED et al. (1958); GIRARDI (1961); MORGAN and LOVERING (1963).

Radiochemical separation is generally required to get rid of all interferences from other activation products. The introduction of high resolution Ge(Li) detectors has offered the possibility of non-destructive analysis of many activation products by their gamma emissions (PRUSSIN et al., 1965). Gamma spectrometry of activation products presents a different situation compared to the passive gamma spectrometry of rocks. A neutron activated source usually contains a large number of high activity emitters with no genetic relationship. The requirement of high resolution is much more important than high efficiency. GORDON et al. (1968) have determined the contents of 24 neutron activated isotopes, including ^{232}Th, in standard rocks by repeated measurements of gamma activity with Ge(Li) detectors. The determination of U by this method is difficult because of the short half life of ^{237}Np and the many interferences in the energy region of intense ^{237}Np gamma emissions.

Alternative approaches can be used for U determinations. One of them is based upon the high cross section for thermal neutron induced fission of ^{235}U (577 barns) as compared to that of other fissionable natural isotopes (e.g., ^{232}Th < 0.0002 barns) and ^{238}U (<0.0005 barns). (See also Fig.79.) U can therefore be determined by separating chemically some of the radioactive fission products and measuring their activity, provided that their fission yield is known or has been determined experimentally. Early determinations based on this method have been made by FACCHINI and ORSONI (1949) and CURIE and FORAGGI (1951). SMALES (1952); SEYFANG and SMALES (1953) and SEYFANG (1955) have based the determination upon the measurement of the activity of the ^{140}Ba–^{140}La group. EBERT et al. (1957, 1959) and KONIG and WANKE (1959) have measured ^{133}Xe and ^{135}Xe.

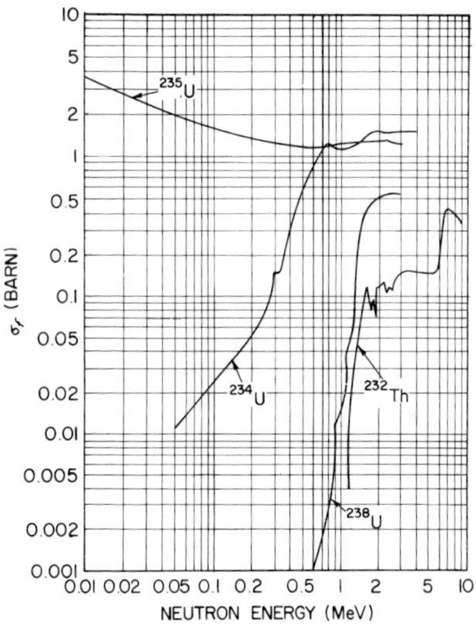

Fig.79. Cross sections (in barns) of neutron induced fissions of ^{234}U, ^{235}U, ^{238}U, and ^{232}Th as a function of the neutrons energy. (After GALE, 1967.)

Another approach for U and Th determinations is based upon the measurement of delayed neutrons emitted by highly excited fission products (ECHO, 1957; AMIEL, 1962; HAMILTON, 1966; AMIEL et al., 1967; GALE, 1967). Some fission products decay by β^- emission to daughter nuclei with energy levels above the respective neutron binding energies. These highly excited nuclei lose their

TABLE XXVI*

THERMAL NEUTRON ACTIVATION DATA OF ^{238}U, ^{235}U, ^{234}U, ^{232}Th, ^{41}K

Nuclide	*Isotopic ratio (%)*	*Thermal neutrons capture cross section (barns)*	*Product nuclei*	*Half-life*	*Decay mode*	*Energy (MeV)*	*Absolute intensity (%)*	*γ Emission energy (MeV)*	*Absolute intensity*
$^{238}_{92}U$	99.276	2.73	$^{239}_{92}U$	23.54 min	β^-	1.29	20	0.044	4%
						1.21	80	0.075	51%
			$^{239}_{93}Np$	2.346 days	β^-	0.713	7		
						0.654	4	0.106	23%
						0.437	48	0.209	4%
						0.393	13	0.228	12%
						0.332	28	0.278	14%
			$^{239}_{94}Pu$	24,390 yr	α	5.16	88	weak	
						5.11	11		
$^{235}_{92}U$	0.7196	101	$^{236}_{92}U$	$2.31 \cdot 10^4$ yr	α	4.49	76		
						4.44	24		
$^{234}_{92}U$	0.0057	95	$^{235}_{92}U$	$7.1 \cdot 10^8$ yr	α	4.55	10.5	0.143	11%
						4.40	84	0.185	54%

TABLE XXVI* (*continued*)

Nuclide	*Isotopic ratio (%)*	*Thermal neutrons capture cross section (barns)*	*Product nuclei*	*Half-life*	*Decay mode*	*Energy (MeV)*	*Absolute intensity (%)*	*γ Emission energy (MeV)*	*Absolute intensity*
$^{232}_{90}Th$	100	7.4	$^{233}_{90}Th$	22.12 min	β^-	1.245	87	0.029	2.1%
						more		0.087	2.7%
								0.453	1%
								more	
			$^{233}_{91}Pa$	27.0 days	β^-	0.568	5	0.087	1.7%
						0.257	58	0.300	6.3%
						0.15	37	0.312	34%
								0.341	3.9%
								0.399	1.1%
								0.416	1.5%
			$^{233}_{92}U$	$1.62 \cdot 10^5$ yr	α	4.82	83	0.042	310 *R*
						4.78	15	0.097	100 *R*
								more	
$^{41}_{19}K$	6.77	1.2	$^{42}_{19}K$	12.36 h	β^-	3.56	82	1.524	18%
						1.97	18		

* Data from LEDERER et al. (1967)—gamma relative intensities (*R*) are reported for ^{233}U.

excess energy by neutron emission rather than by gamma emission. These neutrons are referred to as "delayed" to distinguish them from the "prompt" ones emitted during the fission process. Apart from a number of short lived light nuclei (e.g., ^{9}Li, ^{16}C, and ^{17}N), only the fission produced isotopes lose their excess energy by neutron emission. Neutrons can be selectively counted by the conventional BF_3 detectors and their emission rate is a linear function of the amount of fissionable isotopes present in the irradiated sample.

Fig.79 gives the variation of the fission cross section with the neutron energies (from GALE, 1967). ^{232}Th and ^{238}U are fissioned mainly by neutrons with energies above 0.5 MeV; below this energy only fission of ^{235}U is important. Therefore, if a sample is split into two aliquots, one of them being irradiated by a mixed flux (i.e., both slow and fast neutrons) and the other one by fast neutrons only (slow neutrons being removed, for instance, by a Cd absorber), both U and Th contents can be determined. This technique seems to guarantee a higher accuracy for U compared to Th (GALE, 1967).

Neutron activation techniques have their most useful application for the detection and quantitative estimation of nanograms of U and Th. Uranium concentrations of the order of p.p.b. have been determined in iron meteorites (REED et al., 1958).

Neutron activation techniques for Th and U determinations can be usefully coupled to gamma-ray spectrometry in order to check secular equilibrium conditions in natural systems. It must be pointed out, however, that where induced fission or delayed neutrons techniques are used, ^{238}U determination is based upon the assumption of a constant ^{235}U/^{238}U ratio.

ALPHA SPECTROMETRY

The advantages of alpha over gamma spectrometry of rocks are:

(*1*) Alpha particles from very thin sources lose practically all their energy within the detector, this being due to their very short range. The spectrum of a monoenergetic alpha point source is therefore very simple consisting of one sharp full energy peak.

(*2*) ^{232}Th and ^{238}U, along with other long lived members of the

^{238}U series (e.g., ^{234}U and ^{230}Th), are alpha emitters and very weak gamma emitters. The alpha spectrometry technique therefore offers the possibility of a direct determination of all these long lived nuclei.

(*3*) The background is extremely low and the intrinsic sensitivity very high.

In spite of these essential advantages, alpha spectrometric techniques are used much less than gamma techniques because of the problems arising from the very short range of alpha particles in solids and in liquids. The range of average energy alpha particles in ordinary solids is a few tens of microns and therefore the sample must be deposited in a submicron layer for the self-absorption to be negligible and the spectrum consequently well resolved. The main effect of self absorption on the spectrum shape is the occurrence of strongly asymmetric peaks with more or less pronounced tails on the low energy side (ABROSIMOV and COCHAROV, 1962; VITTOZZI, 1964). The occurrence of prominent tails limits the resolution, this being a serious shortcoming for the determination of long lived natural emitters, whose energies are very close to each other (for instance, ^{238}U and ^{232}Th alpha energies are only about 0.2 MeV apart; and ^{230}Th and ^{234}U are only 0.08 MeV apart). The requirement of an extremely thin layer is therefore imperative in order to resolve clearly the contributions in this crowded part of the spectrum.

The specific alpha activity of natural rocks is very low and the emitters must be radiochemically concentrated in order to get satisfactory counting statistics in a reasonable time. Different extraction techniques for U have been described by GRINDLER (1962) and by LEVINE and LAMANNE (1965). TADDEUCCI et al. (1967) describe extraction techniques for both U and Th. A common procedure is as follows: the sample is dissolved by acid attack (e.g., HF and $HClO_4$) or by fluxes (e.g., sodium hydroxide or sodium peroxide) and the melt is then dissolved in water. U is then extracted by passing the solution through an ion exchange column (CHOPPIN and SIKKELAND, 1960; KORKISH and JANAUER, 1962) or by solvent extraction. For the latter method, U must be converted into uranyl ion, which is readily extracted into organic solvents. Th can be extracted by passing a HCl solution through a cation exchange resin column and by eluting the adsorbed Th with oxalic acid. Complete recovery of U and Th cannot, of course, be guaranteed by any of these procedures and the critical influence of many factors makes the estimate of the recovered fraction very difficult. Quantitative determinations can be carried out by an isotope dilution method, i.e., by adding to the sample solution a known amount of other alpha emitting isotopes. ^{235}U can be used as a spike when only U has to be determined; an artificially prepared ^{228}Th–^{232}U mixture can be used for both U and Th determinations.

The radioactive phase is then deposited on a metal planchet. Common deposition techniques are either evaporation or electrodeposition. The evaporation technique is very simple and quantitative but usually produces non-uniform layers. Electrodeposition gives better results and is to be preferred when very similar alpha energies have to be resolved. U can be electrodeposited from a number of solutions (e.g., acetate, oxalate, formate, fluoride, etc.). Electrodeposition can be performed in a lucite (or other inert material) vertical cylindric cell having a vertical rotating platinum anode and metal cathode planchet on its bottom. Electrodepositions can be made at 80°C with a current intensity of about 0.1 A/cm^2. The deposition is strongly affected by the stirring rate, current intensity, foreign ions, etc.

Painting techniques are sometimes used, giving poorer results than electrodeposition; uranyl nitrate is dissolved in alcohol and added to a dilute solution of cellulose in amyl acetate. The resulting emulsion is painted on the planchet. This method has of course the disadvantage that the deposition layer is diluted by inactive material.

GRAEFFE and NURMIA (1961) have suggested a different approach using a thick source, i.e., a source whose thickness is large compared to the range of all the alpha particles in the source itself.

Special ionization chambers and gold barrier silicon semiconductors are commonly used as detectors. Alpha sensitive scintillators, e.g., cadmium sulphide, can also be used. Ionization chambers with different resolutions and sensitivities have been described by many authors, including LONATI et al. (1958); MCFARLANE and KOHMAN (1961); OGAWA et al. (1961); DOKE (1962); NURMIA (1962); HILL et al. (1964).

AUTORADIOGRAPHIC TECHNIQUES

Autoradiographic techniques are based upon the high sensitivity of photographic emulsions to ionizing radiations from radioactive sources, a phenomenon whose use dates back to the discovery of radioactivity. This method is more suitable for the study of alpha than beta and gamma activities. Alpha particles have in fact very short ranges in photographic emulsions and tracks due to single particles can be resolved and counted. The longer range of beta and gamma radiations makes their resolution a difficult problem. Furthermore a very high sensitivity emulsion is needed in order to compensate for their lower ionizing power.

Emulsions normally used for radioactivity measurement and detection (nuclear emulsions) have a high silver halide content and, in order to increase the resolution, are very fine grained. A number of different types are commercially available from Ilford and Kodak. Table XXVII reports the ranges of alpha particles from natural emitters in Ilford C 2 nuclear emulsions, after DEUTSCH et al. (1956).

Photographic films having a high sensitivity to ultraviolet photons are also commercially available and can be used. Fig.80 shows an autoradiography obtained by a Kodak k k industrial X-ray film. The light grey background is due to photons and β^- rays, the black areas are due to alpha particles and clearly show the sites of alpha emitters. The black stripe corresponds to a thin mineralized vein.

Autoradiographic techniques have been extensively used in the past for evaluating the Th and U contents of minerals, rocks, and meteorites. The usual procedure is to place a polished section of the rock in contact with the emulsion and to let it expose under dark room conditions for some weeks in order to obtain good counting statistics. Determinations are based either on counting of the longest alpha tracks due to ^{214}Po and ^{212}Po (see Table XXVII) or on the evaluation of five prong and three prong stars. A single ^{228}Th nucleus can in fact undergo five subsequent alpha emissions over a 72–720 h exposure time, whereas nuclei from the ^{238}U series can only emit a maximum of three consecutive alpha particles during the same exposure time. Extensive references on the application of

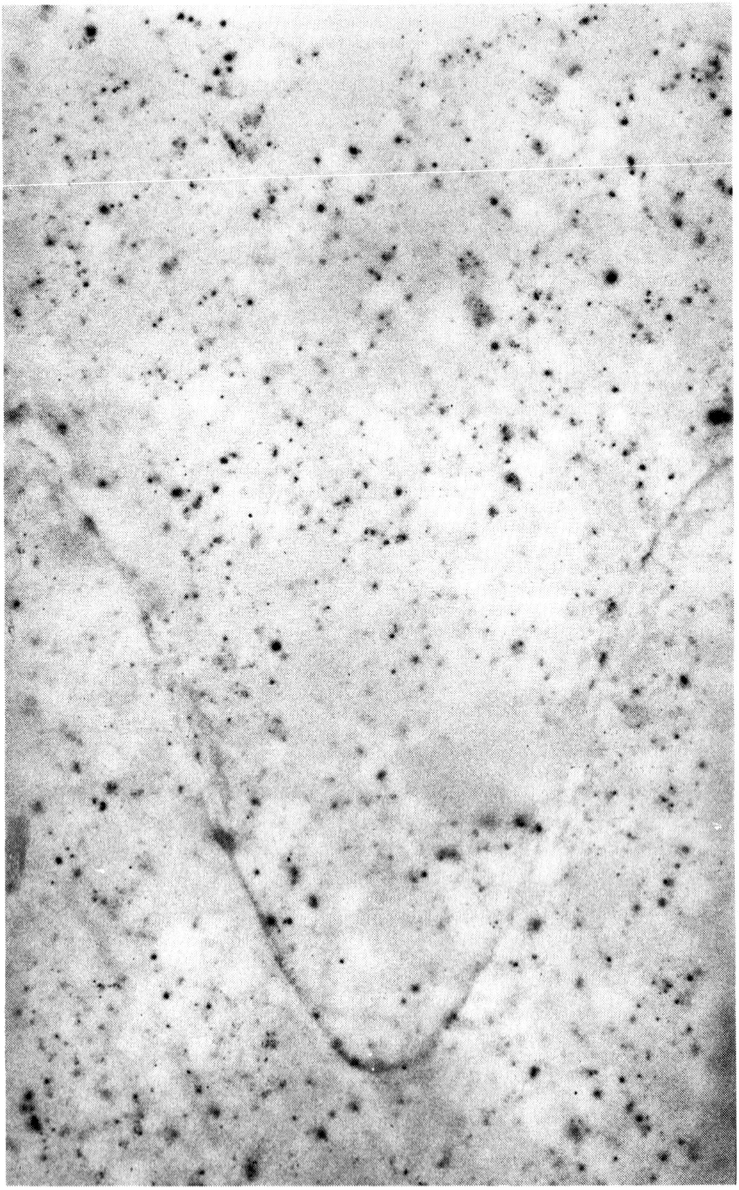

Fig.80. Autoradiography of a core from Conway granite, New Hampshire, U.S.A. (J. A. S. Adams, unpublished data.)

TABLE XXVII

ALPHA PARTICLE RANGES OF ALPHA EMITTERS FROM ^{238}U, ^{232}Th, AND ^{235}U SERIES IN ILFORD C2 NUCLEAR EMULSIONS
(After DEUTSCH et al., 1956)

^{238}U *series*		^{232}Th *series*		^{235}U *series*	
nuclide	*average range* (μ)	*nuclide*	*average range* (μ)	*nuclide*	*average range* (μ)
^{238}U	16.0	^{232}Th	15.0	^{235}U	27.2
^{234}U	19.0	^{228}Th	23.0	^{223}Ra	24.8
^{230}Th	18.9	^{224}Ra	24.7	^{219}Rn	32.4
^{226}Ra	19.5	^{220}Rn	28.7	^{215}Po	36.4
^{222}Rn	23.4	^{216}Po	32.1	^{211}Bi	31.0
^{218}Po	26.8	^{212}Bi	27.6		
^{214}Po	39.0	^{212}Po	47.5		
^{210}Po	22.0				

the method are reported by RANKAMA (1963). Reviews of these techniques and applications have been published by YAGODA (1949), BOWIE (1954), and RAGLAND (1964). All quantitative methods are, however, based upon the assumption of secular equilibrium and of uniform distribution throughout the sample.

The use of autoradiography as a quantitative technique has been obsolescent after the introduction of solid detectors. The main additional shortcoming of the autoradiographic method is the long time needed for an exposure, the tedious work involved, and the difficulty of obtaining satisfactory counting statistics.

Autoradiographic techniques can be very helpful for the identication of the sites of alpha emitters in rocks and minerals and for a quantitative estimate of the relative contribution to the overall alpha activity. The use of liquid emulsion, as described by GUILBERT and ADAMS (1955), is very convenient in this regard. Thin sections of a rock sample are loaded with an emulsion binder solution and dried in a horizontal position. Kodak NTA emulsion, warmed to fluidity in a 45°C water bath, is then applied as uniformly as possible to the binder coated top of the thin section. The emulsion layer is dried and the sample kept over the whole exposure time in a light-tight box at 5°C. The emulsion is then developed, dried and the specimen is examined under the petrographic microscope. By focusing on the emulsion above the thin section, alpha tracks can be studied at 200–450 magnifications with regard to the track density and site. By focusing on the thin section, the mineral grains can be identified by conventional methods. A drawback of the use of liquid emulsion is its fast deterioration. This technique has been applied to the study of the radioactivity distribution in Mount Lassen volcanics (ROGERS and ADAMS, 1957).

An additional application of autoradiography has been the estimation of the relative age of U intrusions in uraniferous coal by studying the relationships between alpha emitter sites and microfaulting (JEDWAB, 1965; HUGI and JEDWAB, 1966).

RADIATION DAMAGE IN SOLIDS

Alpha, beta, and gamma rays passing through solids spend most of their energy in ionization and a small fraction of it in the displacement of ions. Most of the energy of neutrons and recoil nuclei is spent in the dislocation of ions. As a result, the passage of radioactive particles and radiations through solids produces a progressive disorder with consequent alteration of many physical properties, including X-ray diffraction patterns, coloration, hardness, density, indices of refraction, etc.

Naturally radioactive minerals are subject to damage mainly as a consequence of alpha emissions. Many metamict minerals (e.g., zircons, fergusonite, samarskite, etc.) may be so extensively damaged that the internal lattice ordering of ions is essentially destroyed, anisotropy is remarkably reduced, density is abnormally low and other physical properties are significantly altered. Where radioactive elements are unevenly distributed throughout the crystal, complete metamictization may be restricted to highly radioactive zones within the crystal. When metamict minerals are held at suitable elevated temperature, many of the physical properties revert and a definite X-ray diffraction pattern is exhibited, indicating that heating has supplied ions with enough energy for them to return to regular lattice positions. In most cases the restored lattice structures differ from the original ones. The restoration process is an exothermal reaction (a release of 20 cal./g in polycrase minerals and of 65 cal./g in fergusonite has been reported by DANIELS, 1954), indicating that energy storage occurs during metamictization.

Attempts have been made to identify properties that may be quantitatively useful in determining the total intensity of the radiation received by a given crystal since its formation (dose) and for absolute age dating. Thermoluminescence and pleochroic haloes have received the most attention in these regards.

Thermoluminescence is the property, exhibited by many minerals, to emit visible light upon heating. It is believed to be the result of the removal of electrons from lattice imperfections (traps) to normal lattice positions. Natural radiations are the main ionizing agents causing trapping of electrons in lattice imperfections and the total trapped charges in a given mineral can be thought to be a linear function of the irradiation received from a crystal throughout its life. The quantitative evaluation of the thermoluminescence light emitted upon heating can therefore be a measure of the irradiation dose (when the age of the mineral is known) or of the age upon estimate of the irradiation intensity. Many disturbing effects restrict the applicability of the method, such as the presence of various impurities, recrystallization processes, pressure effects. The linearity of the relationship between trapped electrons and total irradiation is further affected by saturation effects, i.e., the linearity holds best where less than one fifth of the total traps are populated.

Artificially thermoluminescent materials, e.g., calcium or lithium fluorides and others, are currently used for radiation dosimetry. Their properties and applications have been reviewed by SPURNY (1965) and FRANCK et al. (1964). Natural thermoluminescence has been reviewed by ZELLER (1954); HOUTERMANS (1961); RANKAMA (1963) and LEVY (1966). ZELLER et al. (1967) have proposed an electron spin resonance method for the measurement of trapped charges in minerals.

Absorption by the host mineral of alpha particles emitted from radioactive inclusions produces a halo of radiation damage surrounding the radioactive impurity. These haloes have been called pleochroic because their color changes while being rotated under cross polarized light. Haloes are frequently observed in natural minerals, mainly in micas, containing small inclusions of zircon or other radioactive minerals. The color intensity of the halo may be a linear function of the alpha total irradiation in favorable cases with small to moderate doses. Many disturbing effects restrict the use of this halo method as a quantitative tool, the most important ones being the different sensitivity of individual minerals to alpha irradiation and the phenomenon of saturation and reversal of the color intensity. Pleochroic haloes methods have been tentatively applied to absolute dating of minerals (DEUTSCH, 1957) but the results have not been encouraging because of the above mentioned shortcomings.

The Russian physicists Flerov and Petrjak discovered in 1940 that ^{238}U nuclei undergo spontaneous fission. Decay constants for spontaneous fission have been determined by a number of authors, the obtained values ranging from $5.3 \cdot 10^{-17}$ to $1.6 \cdot 10^{-16}$ year^{-1} (KAUFHOLD and HERR, 1967). This means that ^{238}U spontaneous fission rate is about a million times lower than alpha emission rate, but is nevertheless an appreciable phenomenon in that about 25 nuclei/h per gram of ordinary U undergo spontaneous fission. Other heavy nuclei, such as ^{235}U and ^{232}Th, exhibit this phenomenon, but with much lower rates (PRICE and WALKER, 1963a).

Fission products have a very high kinetic energy and produce fission recoil tracks that are a few microns long and about 10 mμ wide. Damage tracks can be observed with the electron microscope (SILK and BARNES, 1959; BONFIGLIOLI et al., 1961). The size of the tracks can be increased by a method developed by Fleischer, Price, and Walker (PRICE and WALKER, 1963a; FLEISCHER and PRICE, 1964b; FLEISCHER et al., 1965a). It consists of attacking a clean section of the sample by a proper etching solution which, by etching preferentially the disordered damaged region, increases the size of the tracks which are then visible under the ordinary light microscope. The tracks become visible as hollow narrow cylinders or, after longer etching times, as typical diamond shaped etch pits.

The density of fossil tracks on the etched surface of a sample (ρ_s) is given by:

$$\rho_s = [\exp(\lambda T) - 1]\lambda_{SF} N_0 C_u K/\lambda \qquad (6.1)$$

where: λ, λ_{SF} = ^{238}U alpha emission and spontaneous fission decay constants; T = age of the mineral; N_0 = atomic density of the mineral; C_u = atomic concentration of ^{238}U; K = geometric constant.

Contributions from ^{235}U and ^{232}Th spontaneous fissions and from other causes of fission (cosmic rays, etc.) are negligible for ordinary terrestrial material (PRICE and WALKER, 1963*a*). The dependence of the fossil track density of a sample on its age has possibilities for the applications in geochronology (PRICE and WALKER, 1963a; FLEISCHER and PRICE, 1964a); and its dependence on U concentration for its use as an internal U solid state detector. But, in practice, the evaluation of C_u in eq.(6.1) is complicated by the required knowledge of T. A different approach is therefore used based on the fact that ^{235}U is practically the only natural isotope to undergo thermal neutron induced fission. It consists of etching a sample, counting the fossil tracks, irradiating the sample with thermal neutrons, etching and counting again the tracks. The U concentration (C_u) is given by:

$$C_u = \frac{\rho_{TN}}{N_0 \sigma \varphi I K} \tag{6.2}$$

where ρ_{TN} is the track density due to ^{235}U thermal neutrons induced fissions, σ is the relative cross section, φ is the dose of thermal neutrons, I is the isotopic abundance of ^{235}U and the other symbols are the same as in eq.(6.1). The constant K must be determined experimentally case by case (PRICE and WALKER, 1963b; BERZINA and DOLOMANOVA, 1967). Micas are a most favorable material, and this method allows one to determine contents as low as 4 femtograms, with an adequate neutron flux. Very pure mica can also be used as an external detector by placing it close to a polished section of whatsoever sample and counting the tracks induced in the mineral after irradiation. A difficulty arising from the use of mica sheet as a fission tracks detector is the comparison of the recorded fission tracks with the U bearing phases in the rock. The use of a Lexan plastic sheet as a detector is very helpful for this purpose (KLEEMAN and LOVERING, 1967). In fact, a detailed print of the rock texture forms, after etching, on the surface which was in contact with the sample.

The Th/U ratio can be determined by counting the track production after a thermal neutron and then again after a fast neutron or proton irradiation. The track density production after the first irradiation is:

$$\rho_{\mathrm{TN}} = C_{\mathrm{u}} N_0 \sigma_{\mathrm{UT}} \varphi K \tag{6.3}$$

where σ_{UT} is the cross section for thermal neutrons induced fissions of ordinary uranium, and in the second irradiation:

$$\rho_{\mathrm{FN}} = C_{\mathrm{u}} N_0 \sigma_{\mathrm{UF}} \varphi K' + N_0 C_{\mathrm{Th}} \sigma_{\mathrm{TF}} \varphi_{\mathrm{F}} K'' \tag{6.4}$$

where σ_{UF}, σ_{TF} are the fast particles induced fission cross section for ordinary U and Th, φ_{F} is the fast particles flux, C_{Th} is the atomic concentration of Th in the sample. Since the ranges of fission fragments in the same sample are approximately equal ($K = K' = K''$), by combining eq.(6.4) and (6.3), we have:

$$\frac{C_{\mathrm{Th}}}{C_{\mathrm{u}}} = \left(\frac{\rho_{\mathrm{FN}}}{\rho_{\mathrm{TN}}}\right)\left[\frac{\sigma_{\mathrm{UT}}\varphi}{\sigma_{\mathrm{TF}}\varphi} - \frac{\sigma_{\mathrm{UF}}\varphi_{\mathrm{F}}}{\sigma_{\mathrm{TF}}\varphi_{\mathrm{F}}}\right] \tag{6.5}$$

The problem of the determination of the fast neutron spectrum can be avoided when a sample with known Th/U ratio is irradiated (HUANG et al., 1967). A further interesting application of fission tracks can be the study of the movement of U atoms in crystals by comparing the location of the fossil and of the induced fission tracks (FLEISCHER et al., 1965b). The influence of the etching conditions and of the neutron dose determinations on the precision of the measurement has been discussed by TOGLIATTI (1965a, b) and by KAUFHOLD and HERR (1967).

Fossil tracks produced by recoil nuclei from alpha emission from natural emitters have been identified by HUANG et al. (1967). They appear as a background to the fission diamond shaped etch pits and can be clearly observed under phase contrast conditions. Their number is again proportional to the U and Th content and to the age of the sample, but in this case equilibrium must be assumed in all natural series in order to apply the method for absolute dating. Alpha recoil tracks can give helpful indications about the sites of U and Th impurities in minerals or rocks.

CHEMICAL METHODS

Fluorimetric and colorimetric methods are most used for U determination at p.p.m. levels. Detailed procedures have been described by GRIMALDI et al. (1954) and by ADAMS and MAECK (1954).

Rock samples must be opened, either by hydrofluoric–nitric acid treatment or by any equivalent method, and U separated. U can be extracted by organic solvents (e.g., ethyl acetate) from a solution of high nitrate salt concentration or by nitrated cellulose chromatographic methods. This last method consists of selective elution of uranyl nitrate from activated cellulose with diethyl ether containing 5% by volume of concentrated nitric acid (70%).

Fluorimetric analysis is based upon the property of fluorescence of U when exposed to ultraviolet light. Fluorescence is specifically from U when excited with long wavelength (3,650 Å) ultraviolet light. The quenching effect of many ions (e.g., iron, manganese, cobalt, etc.), present in ordinary rock in a much larger amount than U, makes necessary a highly selective separation of U. The method has been often applied for U determinations in the range 0.05–50 p.p.m. Reported errors are in the range 4–15% in most cases.

Many reagents have been proposed for colorimetric determinations, alkaline peroxides and dibenzoylmethane being mostly used. A serious problem occurring for any reagent is the fact that many other metallic ions develop colors similar to U. Therefore, here again the success of the method is strongly dependent on the effectiveness of the separation. Measurements can be made either by visual comparison with a series of standards in Nessler tubes or by spectrophotometers. The method has been applied for U content in the range 0.2–10 p.p.m. LEVINE and GRIMALDI (1958) have described a chemical method for the determination of Th at p.p.m. level. The method is based upon mesityl oxide selective extraction of Th and spectrophotometric determination. Chemical determinations of Th are expensive in terms of man-hours/determination and many laboratories prefer the gamma spectrometric methods.

Chemical methods of Th and U analysis have been reviewed by GRIMALDI (1961) and BOOMAN and REIN (1962) respectively.

COINCIDENCE COUNTING

Coincidence methods take advantage of the practically contemporaneous emission of two (or more) nuclear particles and/or photons by a given nuclide. According to the nature of the detected emissions, the methods are referred to as α–γ, β–γ, γ–γ, etc.

A coincidence instrument usually consists of two detectors which are often connected through a pulse height selector to a coincidence circuit controlling the entrance of the signals into the pulse height analyzer, or even a simple scaler. The detectors are arranged with geometries that optimize the contemporaneous detection of the coincidence events and minimize spurious effects. The pulse height selectors sort the amplitudes of the pulses reaching the coincidence circuit, preventing thereby the recording of coincident events of any other energies. The arrival of the signals from one of the detectors to the coincidence circuit triggers it so that the coincident event from the other detector enters the coincidence gate and is recorded. A detailed discussion of coincidence instrumentation is given by WAPSTRA (1965). A more complete approach is offered by the development of multidimensional pulse height analyzers which allow a contemporaneous record of coincident and single events from both detectors (see Chapter 3, pp.72–73).

Coincidence measurements of the major radionuclides occurring in rocks do not offer substantial advantages in respect of ordinary gamma spectrometry techniques. Practically all the alpha emitters from the ^{238}U and ^{232}Th series have a very weak gamma activity. ^{226}Ra is the most intense gamma emitter, with about 4% of the alpha disintegrations accompanied by gamma rays (see Table XV). γ–α coincidence techniques are therefore ineffectual. ^{235}U has a comparatively intense gamma emission (about 65% of total disintegration) accompanying alpha emission. Other alpha emitters of this series (i.e., ^{231}Pa, ^{227}Th, ^{223}Ra, ^{211}Bi) have also enough intense coincident gamma emission. α–γ coincidence measurements could be in principle applied to the detection of ^{235}U in U and Th minerals, provided that the contribution by ^{226}Ra is negligible or can be corrected for. Because of the high self absorption of alpha particles, this technique can only be applied to U rich ores or minerals.

At least four β^- emitters exist in the ^{232}Th and ^{238}U series (i.e., ^{212}Pb, ^{208}Tl, ^{214}Pb, and ^{214}Bi) which are also intense gamma emitters. The decay schemes of all four nuclides are not simple and β–γ techniques are not very discriminative.

A method of β–γ counting has been proposed (see EICHHOLZ et al., 1953), which allows one to determine the U content irrespective of the equilibrium conditions in the ^{238}U series. This method is based upon a simultaneous measurement of the gross β^- and gamma activity of the sample. The gamma activity comes essentially from ^{226}Ra daughters, whereas the combined β^- activity of the ^{226}Ra daughters is slightly higher than that of ^{234}Th and ^{234}Pa. A comparison of gross β^- and gamma activity allows one to evaluate the real U content of the sample.

PROSPERI and SCIUTI (1958) have described a method of γ–γ–β coincidence counting, which allows one to discriminate the contributions from the Th series from those of the U series, even if the ^{238}U series is not in equilibrium. This method is based upon the gross counting of β^- particles (e.g., by a GM counter) and gamma rays (e.g., by a scintillation crystal). The pulses coming from the scintillation crystals are sent to two pulse height discriminators, one allowing the counting of pulses from both ^{234}Th and ^{226}Ra daughters, and the other allowing the counting of pulses from ^{226}Ra daughters only. The method allows one to evaluate three unknowns (U content, Th content, and disequilibrium degree in the ^{238}U series) by a system of three equations relating these unknowns to the gross β^- counting and to the two gamma countings.

γ–γ coincidence counting is possible whenever a nuclide does not give off all its excess energy by one gamma emission, but passes through intermediate energy levels by multiple gamma emission (gamma rays in cascade). ^{208}Tl is a major natural radionuclide having the emission of gamma rays in cascade in 100% of the disintegrations (see Fig.57). Practically all the decays leave the ^{208}Pb nucleus in a highly excited state; all of which decay mostly by gamma emission from the 2.615 MeV excited level. Each gamma transition is therefore accompanied by a coincident 2.615 MeV gamma ray. Coincidence γ–γ measurements of ^{208}Tl are easily made, but they do not offer any advantage in respect to ordinary gamma spectrometry in that the ^{208}Tl 2.615 MeV emission is practically the most energetic in nature, so that ordinary gamma spectrometry techniques discriminate it very clearly from any other natural activity.

γ–γ coincidence methods have a useful application in the detection of minute amounts of some cosmogenic nuclides, such as ^{22}Na and ^{26}Al. Both of these isotopes decay mostly by positron emission (^{26}Al = 85%; ^{22}Na = 90%), accompanied in all cases by a gamma emission (^{26}Al = 1.81 MeV; ^{22}Na = 1.275 MeV). The positron annihilation produces two coincident 0.511 MeV gamma rays emitted at nearly 180°. Three coincident gamma rays are therefore emitted for each positron decay. If two NaI(Tl) crystals, or other equivalent gamma detectors, are placed on the opposite sides of the source and the signal coming from the absorption of a 0.511 MeV gamma ray by a preset detector is chosen to trigger the coincidence circuit, two coincidence peaks will be recorded at 0.551 MeV and at 1.81 MeV (for ^{26}Al) or at 1.275 MeV (for ^{22}Na). The method is highly discriminative against all the other nuclides, except positron emitters and a few nuclides with very complex decay schemes. This method has been applied to the detection of ^{22}Na and ^{26}Al in meteorites and of ^{26}Al in deep sea sediments by many authors, including ANDERS (1960); LIPSCHUTZ et al. (1965); WASSON et al. (1967); HERPERS et al. (1967).

CHAPTER 7

Case Histories

GEOCHRONOLOGICAL

The determinations of the abundances of natural radioisotopes and their stable end products in a rock or mineral can be used in favorable cases to estimate the time during which the stable radiogenic isotope accumulated in the system. Each of the natural radioactive series can be used potentially as a "radioactive clock".

Mass spectrometry techniques are usually used for the determination of both parent and end products. In the ^{40}K–^{40}Ar method, gamma spectrometric determinations of both isotopes can be carried out with some accuracy and precision for dating purposes. ^{40}K can be determined by passive gamma-ray spectrometry upon its 1.46 MeV gamma emission (see Chapter 4). ^{40}Ar can be determined only after transmutation to ^{41}Ar upon activation by thermal neutrons.

Ar analysis by neutron activation

The usual method of Ar analysis is by mass spectrometric isotope dilution using an ^{38}Ar (or ^{39}Ar) spike. The non-radiogenic isotope ^{36}Ar (natural abundance = 0.337%) is used for evaluating the eventual fraction of ^{40}Ar due to atmospheric contamination (natural ^{40}Ar/^{36}Ar ratio = 295.6).

^{36}Ar and ^{40}Ar can also be determined upon activation by thermal neutrons and subsequent measurement of the activities of ^{37}Ar and ^{41}Ar produced by the reactions:

$$^{36}\text{Ar}\ (n, \gamma)\ ^{37}\text{Ar} \qquad \text{(cross section: 6 barn)}$$

$$^{40}\text{Ar}\ (n, \gamma)\ ^{41}\text{Ar} \qquad \text{(cross section: 0.61 barn)}$$

TABLE XXVIII

RADIOACTIVE CONSTANTS OF ^{37}Ar AND ^{41}Ar
(After LEDERER et al., 1967)

Isotope	*Half-life*	*Mode of decay*	E_{β^-}*(MeV)*	E_{γ}*(MeV)*	*Abs. int.* (%)
^{37}Ar	35.1 days	E.C.	–	–	–
^{41}Ar	1.83 h	β^-	1.20, 2.49	1.293	99.0

The radioactive constants of ^{37}Ar and ^{41}Ar are reported in Table XXVIII.

The analytical procedure can be summarized as follows: Ar is extracted by fusion under high vacuum and is subsequently purified from extraneous gases. Purified Ar is transferred into a high purity silica vial and is irradiated, along with a standard of known isotopic composition, by thermal neutrons. Fluxes of the order of 10^{12}–10^{14} neutrons/cm^{2}sec are generally used. Because of the different half-lives of ^{37}Ar and ^{41}Ar, different irradiation times are required in order to obtain detectable quantities of the two isotopes. An irradiation time of several hours is generally adequate for ^{41}Ar, but much longer irradiation times are needed for ^{37}Ar. After irradiation, Ar is again purified in order to get rid of long lived interferring activities and it is then transferred to a gas filled detector. Proportional counters filled with Ar and methane are commonly used. Gross counts of sample and standard are carried out at discrete time intervals, and ^{41}Ar and ^{37}Ar are determined from the observed decay curves. This method has been used with minor modifications by many authors, including MOLJK et al. (1955); STOENNER and ZAHRINGER (1958); WANKE and KONIG (1959); SOROIOU et al. (1965) and ARMSTRONG (1966). Although not as accurate, precise, and sensitive as isotope dilution, this technique has the appeal of comparatively cheap instrumentation, where reactor facilities are available.

As shown in Table XXVIII, ^{41}Ar is an intense gamma emitter. As the gamma emission is monoenergetic and its energy is comparatively high, gamma spectrometry is an appropriate and very simple tool for its measurement. High resolution is not generally needed and well-type NaI(Tl) crystals can be consequently used in order to increase the total efficiency. Absolute calibration can be carried out with ^{22}Na 1.275 MeV and ^{60}Co 1.332 MeV gamma standard sources. It must be recalled that when an isotope emitting gamma rays in cascade is counted, corrections must be made for coincident events resulting from the nearly simultaneous absorption of cascade gamma emissions.

Gamma spectrometric determinations can be quite effective in discriminating against disturbing impurities with half-lives close to that of ^{41}Ar. An example of interfering activity of this kind has been reported by KAISER and ZAHRINGER (1968). They have measured by total counting an abnormally high activity with a half-life corresponding to that of ^{41}Ar in the troilite phase of a meteorite. This was found to be due to the build-up of ^{83m}Kr from ^{82}Se. ^{83m}Kr has in fact a half-life of 1.83 h. Chemical separation of Kr from Ar was therefore necessary to get rid of the contribution from ^{83m}Kr to the total counting. ^{83m}Kr has a very low gamma energy and also internally converts most of its excess energy. Gamma spectrometry would therefore permit an easy discrimination of the activities due to these isotopes with equal half-lives.

The counting times needed for obtaining satisfactory counting statistics is of course dependent on many factors, including the amount, the age and the K content of the sample, size of the crystal, irradiation flux and time, etc.

^{37}Ar does not emit any particle or gamma radiation. It must be determined therefore with the X-rays emitted after electron capture decay. Argon and methane filled proportional counters are the most suitable detectors for this purpose. Semiconductors, detectors, e.g., Si(Li), can also be used, but they suffer a much poorer geometry.

Determination of ^{40}K

^{40}K is usually evaluated by determining elemental K by chemical, flame photometric, or X-ray fluorescence techniques. Neutron activation (see Chapter 6, pp.167–168) or mass spectrometric isotope dilution are also used in some cases (particularly with very low K contents). These last techniques, however, determine one of the more abundant isotopes, namely, ^{39}K or ^{41}K. Therefore ^{40}K can be evaluated by any of these techniques only if a constant isotopic composition of K in nature is assumed. Some cases of isotopic fractionation of K in geologic processes have been reported, the fractionations found to date can only cause negligible errors (~ 1–2%).

A direct determination of ^{40}K can be carried out by gamma-ray spectrometry. The determination is based upon the measurement of the intensity of the 1.46 MeV gamma emission and offers the following substantial advantages: (*1*) no assumption about the isotopic composition of K is required; (*2*) the analysis is non-destructive and no alteration of any kind is induced in the sample. Ar can, therefore, be determined on exactly the same aliquot used for K determination. Moreover, as all the operations consist merely of weighing, loading, and spectral analysis, this method is time-saving for the operator. The long counting times on the spectrometer for a single measurement may be a drawback where an extensive program of radiometric analyses is in progress. One or two determinations per day can be maintained with most instrumentation where at least a few grams of sample with K content higher than 1% are available.

The ultimate limits of sensitivity and precision of the method are strictly dependent on the characteristics of the instrumentation and on the available amount of sample. CIVETTA et al. (1969) have investigated the reliability of routine gamma spectrometric determinations carried out on a few grams of mineral sample with K contents in the range 0.5–17%. Absolute calibration of this quite ordinary spectrometer was carried out by salts where K occurs with the natural isotopic composition and by artificial radioisotopes mixed in a substance having density and atomic number close to those of the sample. The negligible variability of the ^{40}K isotopic ratio in K salts was checked by comparing results obtained with different salts. ^{60}Co (1.17 and 1.33 MeV) and ^{88}Y (1.81 MeV) were the artificial radioisotopes used for absolute calibration.

Table XXIX reports the results of determinations carried out on salts having different density and average atomic number. Three different plastic cylindric canisters were used. They had the same base diameter (3.5 cm) and were 0.5, 3.0, and 5.0 cm high. They were completely filled with the salt and placed on the detector. All the analyses have been carried out at the Istituto di Fisica Terrestre of the Università di Napoli, described in Chapter 3, pp.84–85.

TABLE XXIX

K FULL ENERGY SPECIFIC ACTIVITY RECORDED WITH VARIOUS SALTS WITHIN CYLINDER CANISTERS WITH THE SAME DIAMETER AND DIFFERENT HEIGHTS

Salt	*Density in the container* (g/cm^3)	*Height of the container* (*cm*)	*Recorded pulses/min g K in the full energy peak*
K_2CO_3	1.03	0.5	5.62 (±0.10)
	1.03	3.0	4.74 (±0.07)
	1.03	5.0	4.32 (±0.06)
KNO_3	1.37	0.5	5.56 (±0.10)
	1.37	3.0	4.80 (±0.06)
	1.37	5.0	4.38 (±0.06)
KCl	1.42	0.5	5.60 (±0.10)
	1.42	3.0	4.86 (±0.08)
	1.42	5.0	4.34 (±0.06)
K_2CrO_4	1.51	0.5	5.52 (±0.10)
	1.51	3.0	4.80 (±0.06)
	1.51	5.0	4.32 (±0.06)

When the specific activities of K measured for different salts in the same container are compared, no systematic variation with the density can be observed. The maximum variation for a 30% change of density is 2%. The lack of any sensitive variation of specific activity with the density is, of course, a consequence of the comparatively high energy of gamma rays which minimizes differential effects of self absorption from the same thicknesses of salts with different density. The specific activity of any salt decreases systematically with the height of the container. This is due mostly to the poorer geometry and to an increasing probability of self absorption for sources of emission more and more distant from the detector. In conclusion, differences of density between sample and standard do not seem to be a major cause of error in K determination, when a few cm thick sources at the same distance from the detector and of the same height are used.

A major cause of uncertainty may be the correction for contributions from ^{214}Bi and ^{208}Tl. BUNKER and BUSH (1967) suggest a graphical correction; they evaluate the amplitude of the 1.46 MeV peak from a line joining the two minima neighboring the 1.46 MeV peak. An examination of the ^{226}Ra spectrum (see Appendix 2) shows that this method is only an approximation. The ^{226}Ra spectrum has in fact a full energy peak at 1.43 MeV (^{214}Bi), which is ignored by the graphical method. The stripping factor can be more correctly evaluated by just summing up the counting rate of the ^{226}Ra shape standard in the ^{40}K channels. The graphical method can be used where the $^{226}Ra/^{40}K$ ratio in the sample is very low. In order to increase the counting statistics all the pulses recorded within the full energy peak should be considered. Where one deals with low activities, it can be expedient to utilize also the part of the spectrum below 1.46 MeV full energy peak. In this case, however, the effect of small differences in densities and atomic numbers is enhanced.

Table XXX reports the average K content ($\overline{K}$) from repeated measurements on samples of muscovite, sanidine, leucite, augite and whole rock. These results show that K contents of 6–8% can be determined in 1–2 g of sample with a precision of 2.5–3%. This precision is less than that obtained by classical methods, but is still adequate for some dating purposes. The precision can, of course,

TABLE XXX

RESULTS OF K DETERMINATIONS (AT LEAST 10 MEASUREMENTS FOR EACH SAMPLE) ON DIFFERENT AMOUNTS OF MINERAL SEPARATES AND ONE WHOLE ROCK SAMPLE FROM ITALIAN VOLCANOES

Mineral	*g*	$\overline{K}$ (%)	2σ	$2\sigma/\overline{K}$ (%)	*Active counting time* (sec)
Sanidine	1.50	7.85	0.20	2.5	65,536
Leucite	2.17	14.94	0.42	2.8	65,536
	11.41	14.85	0.25	1.7	32,768
	20.36	14.89	0.07	0.5	32,768
Muscovite	1.02	7.75	0.20	2.6	65,536
	3.72	7.79	0.12	1.5	32,768
	6.54	7.84	0.10	1.3	32,768
Augite	3.10	0.48	0.11	28.9	65,536
	16.29	0.50	0.08	16.0	65,536
	24.05	0.53	0.03	5.7	65,536
Leucitite (whole rock)	2.50	6.81	0.19	2.7	65,536

be improved still more by using larger samples, or longer counting times, or more sensitive instrumentation. The results obtained on the augite samples show that with K contents of about 0.5%, an acceptable precision can be obtained, only if some tens of grams of sample are used with counting times less than one day.

The accuracy of gamma spectrometric determinations can be estimated by a comparison with results obtained by other independent methods. Table XXXI reports the K determinations made by flame atomic absorption and gamma-ray spectrometry on

TABLE XXXI

COMPARISON OF K DETERMINATIONS BY GAMMA-RAY SPECTROMETRY AND FLAME ATOMIC ABSORPTION ON MINERAL AND WHOLE ROCK SAMPLES FROM VOLCANOES OF SOUTHERN ITALY
(After GASPARINI and ADAMS, 1969)

Formation	*Mineral*	K(%)		*Difference* (%)
		gamma spectr.	*atomic abs.*	
Acquacetosa flow (Colli Albani)	whole rock	7.45 (±0.04)	7.45 (±0.04)	0
Capo Bove flow (Colli Albani)	whole rock	7.18 (±0.22)	6.92 (±0.14)	3.7
Capo Bove flow (Colli Albani)	leucite	16.18 (±0.10)	16.03 (±0.10)	0.9
S. Maria flow (Roccamonfina)	leucite	11.02 (±0.21)	11.21 (±0.22)	1.7
S. Maria flow (Roccamonfina)	whole rock	5.48 (±0.10)	5.61 (±0.16)	2.3
S. Carlo flow (Roccamonfina)	leucite	17.68 (±0.30)	17.14 (±0.22)	2.5
S. Carlo flow (Roccamonfina)	whole rock	8.58 (±0.16)	8.36 (±0.06)	2.6
Mt. Lattani (Roccamonfina)	whole rock	4.01 (±0.06)	4.06 (±0.05)	1.2
Green tuff (Ischia)	biotite	4.30 (±0.10)	4.41 (±0.09)	2.5
Green tuff (Ischia)	sanidine	8.70 (±0.05)	8.66 (±0.01)	0.5
Mt. Vico (Ischia)	feldspars	6.28 (±0.10)	6.33 (±0.10)	0.8
Mt. Vico (Ischia)	whole rock	5.82 (±0.10)	5.86 (±0.04)	0.7
Piperno (Campi Flegrei)	sanidine	8.42 (±0.10)	8.37 (±0.09)	0.6
Piperno (Campi Flegrei)	black part	6.47 (±0.10)	6.34 (±0.02)	2.0
Trachytic lavas from Somma–Vesuvio from a drill	whole rock	4.55 (±0.10)	4.71 (±0.04)	3.5

mineral and whole rock samples (GASPARINI and ADAMS, 1969). A few grams of mineral samples and 15–20 g of whole rock samples were used for the gamma-ray spectrometry. The measurements were carried out at the Department of Geology of Rice University, Houston, by the improved geometry arrangement described in Chapter 3, pp.83–84. Counting times varied from 12 to 20 hours according to the sample counting rate/background counting rate ratio. In all cases the accuracy found was better than 3.7%.

In conclusion, gamma spectrometric determinations of K in ordinary geologic samples can be carried out with gamma spectrometers with a precision and accuracy that, although generally lower than that obtainable with classical physical and chemical methods, are often nevertheless satisfactory for K–Ar work. With anticoincidence shielding still better accuracy and precision should be obtainable.

Determination of the gamma emission rate from natural K

The decay scheme of ^{40}K is reported in Fig.55. As the decay directly to the ground state of ^{40}Ar is practically negligible, the determination of the specific gamma emission rate of ^{40}K is a measurement of the rate of decay of ^{40}K by electron capture.

An accurate and precise determination of the electron capture decay constant (λ_ε) is highly desirable for K–Ar dating. The K–Ar age (t) of a rock or mineral is in fact given by:

$$t = \frac{1}{(\lambda_\beta + \lambda_\varepsilon)} \ln\left(1 + \frac{\lambda_\beta + \lambda_\varepsilon}{\lambda_\varepsilon} \frac{^{40}\mathrm{Ar}}{^{40}\mathrm{K}}\right) \tag{7.1}$$

where λ_β is the decay constant of β^- decay and ^{40}Ar and ^{40}K are the concentrations of ^{40}Ar and ^{40}K in the sample. As discussed by WETHERILL et al. (1956b), t is much more sensitive to an erroneous value of λ_ε than to a comparable error in λ_β. Because λ_ε is of the order of $0.1\lambda_\beta$, t is sensitive to λ_ε only where it does not occur summed with λ_β, i.e., in the product $\lambda_\varepsilon{}^{40}$K of eq.(7.1). If the gamma emission rate of natural K is determined, $\lambda_\varepsilon{}^{40}$K will be independent of the knowledge of the natural isotopic concentration of ^{40}K.

Many of the recent determinations of the gamma emission rate of natural potassium have been carried out by gamma-ray spectrometry. The methods used by WETHERILL (1957) and by LEUTZ et al. (1965) will be here briefly described as an example. WETHERILL (1957) prepared a solution of reagent-grade potassium acetate in water. The weight of potassium acetate was 1.198 kg and the total weight of the solution was 1.694 kg. K concentration was checked by gravimetric and mass spectrometric isotopic dilution techniques. Absolute calibration was carried out by ^{60}Co and ^{24}Na and extrapolating the straight line to 1.46 MeV. A fraction of the solution (242 g) was poured into a polyethylene counting bottle, which was held about 5 cm over the detector (a 3 × 3 inch sodium iodide, thallium activated, crystal) by a polystyrene support. The purpose of this was to reduce the absorption of coincident photons from the 3 × 3 inches NaI(Tl) crystal. Shielding was furnished by 7.5 cm of lead. Background was measured using water as a blank in the counting bottle. Only the pulses recorded within the full energy peak were considered. The specific activity of 3.39 gammas/sec g of K was obtained, with an uncertainty of $\pm 3.6\%$. LEUTZ et al. (1965) have applied two different methods for the evaluation of the specific activity of ^{40}K: (*1*) the β^- and the gamma specific activity has been determined by using 43 KI(Tl) crystals from 0.5 to 500 g in weight; and (*2*) the specific gamma activity has been also measured by five well-type NaI(Tl) and CsI(Tl) crystals, using a KCl source electromagnetically enriched to 30% ^{40}K Cl. KI(Tl) crystals detected β^- as well as gamma rays. Absolute efficiency for β^- rays was practically one, since all β^- particles lose at least a fraction of their energy in the crystal. Efficiencies for gamma rays were evaluated theoretically. Where the total counting rates obtained with different sized KI(Tl) crystals were plotted against the respective gamma absolute efficiencies, a straight line was obtained. The intersection of this straight line with gamma absolute efficiency zero and one gives the β^- specific activity and the gamma plus β^- specific activity respectively. The ^{40}K Cl enriched source was measured in well-known geometry conditions and PVC, tantalium and silver absorbers were used to reduce contributions from external bremsstrahlung and K fluorescence X-rays. The inferred gamma specific activity was 3.25 ($\pm$0.07) gammas/sec g K. A β^- specific activity of 28.26 ($\pm$0.05) β^-/sec g K has been also deduced.

GEOLOGICAL TRACERS

Magmatic differentiation

The petrochemical evolution trends observed in a single igneous suite and the genetic relationships among different sequences are classically discussed in terms of the bulk chemical composition and the phase equilibria. Many determinations of elements contained in trace amounts have also been carried out in recent years. Many trace elements have been observed to have highly variable concentrations in different rock types. The study of their distribution in well known suites has furthermore shown that they behave typically in different evolution series. Sensitive variations of some trace elements (including U and Th) have also been observed in rocks having the same bulk composition, but different magmatic history. All these findings have therefore shown that relative and absolute abundances of trace elements in rocks contain much useful petrologic information and increasing interest is being devoted to their application to the study of magmas.

Although many elements can be determined by gamma spectrometry following activation by thermal neutrons (see GORDON et al., 1968), U and Th are the only trace elements whose concentrations can be directly determined by measurement of their natural radioactivity. Their importance in terrestrial heat production, the wide range of variation (within more than two orders of magnitude) in different rock types, the sensitivity to various magmatic differentiation processes and the economic importance are factors enhancing the interest in the study of their distribution in rocks. The main geochemical characteristics dictating the behavior of U and Th in crystallogenetic processes can be briefly summarized as follows (for a more detailed discussion see ADAMS et al., 1959; ROGERS and ADAMS, 1969a,b):

(*1*) in reducing environments, U and Th are both stable in the tetravalent ion state. Tetravalent Th and U have similar ionic radii, equal coordination number in respect to oxygen (8) and complete outermost electronic shells. Consequently they strongly tend to remain together in geologic processes occurring under reducing conditions. The average Th/U ratio of meteorites and terrestrial rocks is 3–4. Because of their low abundances, quite unusual atomic sizes, high valence and electronegativity, both actinides cannot form isomorphic series involving major rock forming minerals and tend to be segregated in very minor mineral phases. As will be shown below in some examples, both Th and U tend to concentrate in the residual melt during fractional crystallization and are therefore preferentially incorporated in latest stage acidic differentiates;

(*2*) in oxydizing environments, in which Th remains tetravalent, U is oxidized to the hexavalent state. From a geochemical standpoint, the most significant characteristic of compounds of hexavalent U is their comparatively higher solubility in aqueous solutions. As a consequence, hexavalent U is mobilized much more easily than Th. High Th/U ratios in igneous rocks may be therefore indicative of oxidizing conditions experienced by the magma material before crystallization and/or extensive leaching during the postcrystallization history of the rock.

Some of the many examples of the distribution of U and Th in suites of rocks well studied from a petrologic point of view will be described. The examples are the Skaergaard intrusion of Greenland, the Duluth lopolith of Minnesota, and the Campanian volcanic area of southern Italy.

The Skaergaard intrusion of Greenland is considered a typical example of differentiation of a tholeiitic magma by fractional crystallization and gravitational settling. Its surface outcrop covers approximately 60 km^2 and a section of about 3,000 m is at present exposed. The accessible thickness is estimated to represent about 40% of the total. The composition both of each mineral phase and of the whole rock changes gradually upwards. The upward enrichment is towards the low temperature end members of each solid solution series and more acidic bulk composition of the whole rock. A chilled fine grained gabbro with marked flow structures, which outcrops at the border of the intrusion, is believed to represent the composition of the undifferentiated magma. Small volumes of hedenbergite granophyre outcrop at the top of the intrusion. They were originally believed to represent the latest stage acidic differentiates squeezed away by the almost completely solidified mass. $^{87}Sr/^{86}Sr$ ratios and lead isotopic composition indicate, however, that the granophyres are not wholly comagmatic with the whole intrusion. The high $^{87}Sr/^{86}Sr$ ratio (0.707–0.727) would suggest crustal contamination (HAMILTON, 1968).

HAMILTON (1959) investigated the U distribution within the whole Skaergaard sequence. U concentration in whole rock samples was determined by neutron activation and fluorimetry. The radioactivity of the minerals was investigated by autoradiography. The marginal chilled gabbro has a U content of 0.2–0.3 p.p.m. which compares well with that of typical oceanic tholeiites. U increases upward in the intrusion from about 0.1 up to 0.45 p.p.m. at a height of 1,800 m. There it slightly decreases and shows a rather irregular behavior, probably as a result of surfacial weathering. The granophyres have U contents in the range 1.0–3.5 p.p.m. which compares well with the average crustal abundances. A linear relationship between U and K is evident. As concerns minerals, most of the activity comes from accessories, such as zircon, sphene, and epidote. No significant changes have been observed in individual major minerals during differentiation. Where quartz occurs, it becomes the most active major mineral. Acidic feldspars have the activity mostly concentrated along recognizable inclusions, whereas basic feldspars have mostly point sources within the crystal.

The general trends observed at the Skaergaard intrusion have been confirmed by the study of the U and Th distribution at the Duluth lopolith of Minnesota (HEIER and ROGERS, 1963), using gamma spectrometry. This stock appears to have differentiated along the same trend as the Skaergaard intrusion. U contents are on the average comparable with those of Skaergaard rocks, ranging from 0.1 to 5.6 p.p.m. The Th/U ratio increases from 3.3–3.5 (gabbros and ferrogranodiorites) to 4.0 (granophyres). Th/K ratio shows a tendency to increase during differentiation. Also U/K ratio seems to show the same trend, but less markedly.

The Campania volcanic area of Italy is particularly favorable for the study of the behavior of Th and U in differentiation processes other than the simple fractional crystallization. Four volcanic complexes (Roccamonfina, Ischia, Campi Flegrei, and Somma–Vesuvio) with marked individual characteristics are in fact situated in a rather small area (about 4,000 km^2). The absence of any recognizable relationship of activity among the four volcanoes, the very shallow foci of earthquakes related to volcanic activity, the nearly contemporaneous emission of products with different petrochemical composition (e.g., leucitites at Somma–Vesuvio, latites at Ischia and trachytic phonolites at Campi Flegrei) strongly support the hypothesis of emplacement of separated magma reservoirs feeding each volcano and differentiated along individual patterns. The similar K–Ar ages (0.2–0.3 million years) of the oldest accessible products from Somma–Vesuvio and Campi Flegrei and of the revivals of activity at Roccamonfina (<0.3 million years) and Ischia (GASPARINI and ADAMS, 1969) along with the similar petrographic nature of their products of this age suggest a common parent magma. The subsequent evolution pattern of each magma reservoir is probably affected by very local conditions. $^{87}Sr/^{86}Sr$ ratios (0.7071–0.7098) and Rb/Sr ratios (0.25–3.9) suggest an origin at least partially anatectic for the parent magma (HURLEY et al., 1966).

U and Th have been determined on lavas from Somma–Vesuvio (LUONGO and RAPOLLA, 1964), Ischia (GASPARINI, 1963), and Roccamonfina (CIVETTA et al., 1965) by gamma-ray spectrometry. Because of the very young ages of these products, U has been determined with the 0.09 MeV peak (see Chapter 5). The very high extremes reached from both actinides (U up to 43 p.p.m. and Th up to 109 p.p.m.) support further the hypothesis of a crustal contamination of these magmas (IMBO' et al., 1968). Moreover, the occurrence of high K contents confirms the well known tendency of U and Th to be concentrated along with K.

The evolutionary trends at Somma–Vesuvio and Ischia are sufficiently well known so that well grounded hypotheses can be put forward for the relative differentiation processes.

The following rock suite has been erupted during the revival of activity at the island of Ischia: latites, trachytes, alkali-trachytes and sodalite-phonolites. Its genesis has been explained by RITTMANN (1948) by a complex differentiation, involving gravitational settling and gaseous transfer, of a trachytic magma. Fig.81 reports the Th–U diagram of Ischian lavas. A sharp increase of Th and U occurs where gaseous transfer becomes predominant (alkali trachytes and sodalite phonolites). The high extremes reached by the Th/U ratio in such rocks seems to indicate that Th is more sensitive

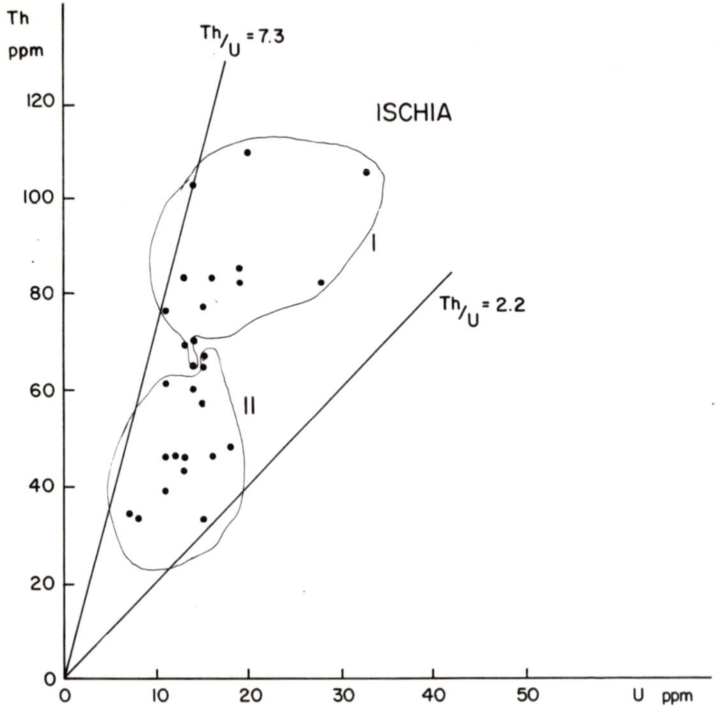

Fig.81. Th–U plot of Ischian lavas: *I* = phonolites, alkali trachytes; *II* = trachytes, latites. (Data from GASPARINI, 1963.)

than U to this process. A characteristic differentiating a suite where gaseous transfer phenomena have played an important role is that there is no direct dependence of Th (or U) on Si. Such kinds of rocks are in fact generally undersaturated because of the high increase of alkalis. As U and Th increase along with alkalis, their abundances increase with increasing undersaturation of the rock.

It is well known that Somma–Vesuvio underwent a magmatic evolution from trachytes to phonolitic leucite tephrites, leucite tephrites, tephritic leucitites, explained by RITTMANN (1933) with the hypothesis of assimilation of Triassic dolomites from the trachytic magma intruded into such rocks. Gravitational settling of mafic crystals and gaseous transfer have had an important part also. The validity of this model has been recently questioned by HURLEY et al. (1966) and by SAVELLI (1967a,b). Average Th and U abundances (Fig.82) decrease from trachytes to leucite tephrites and increase from leucite tephrites to leucitites. The decrease is consistent with the hypothesis of assimilation of low radioactivity

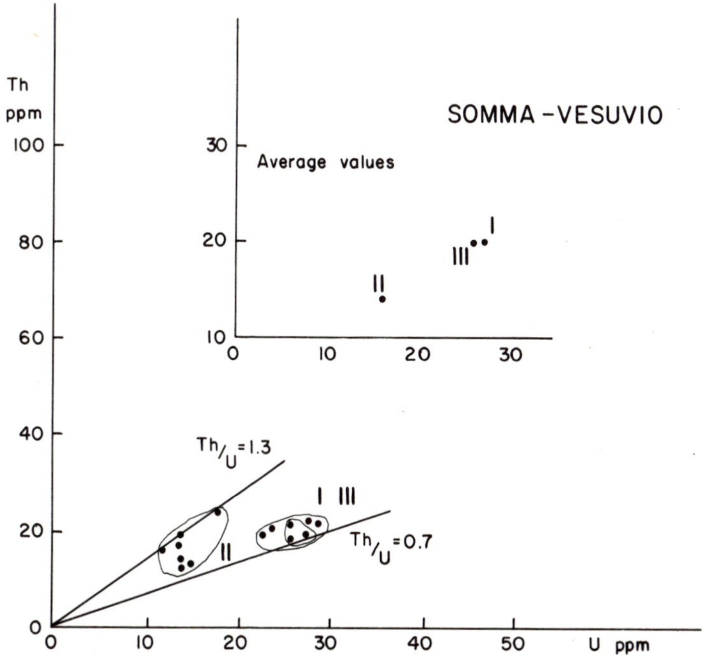

Fig.82. Th–U plot of lavas from Somma–Vesuvio. *I* = Trachytic lavas; *II* = leucite tephrites; *III* = tephritic leucitites. (Data from LUONGO and RAPOLLA, 1964.)

rocks, such as the Triassic dolomites of southern Italy. The subsequent increase is clearly contradicting. If one remains within the model proposed by Rittmann, the increase could be explained by a dominant contribution of gaseous transfer processes at this stage of evolution. This would, in fact, justify also the sharp increase of alkalis. U and Th distribution in the Somma–Vesuvio suite does not contradict the model proposed by Rittmann.

The activity of the Roccamonfina volcano can be schematically divided into two cycles: leucite bearing rocks were erupted during the first one and trachytes to basalts during the second. According to AREVALO-CARRETTERO et al. (1962) the second suite is formed by gravitational differentiation of a new magma. K–Ar ages support this hypothesis. Furthermore, K–Ar ages of leucite bearing rocks indicate that they are distinctly older (1.2–0.3 million years) than the analogous products from Somma–Vesuvio (less than 0.2 million years). Th and U abundances fit well the available data.

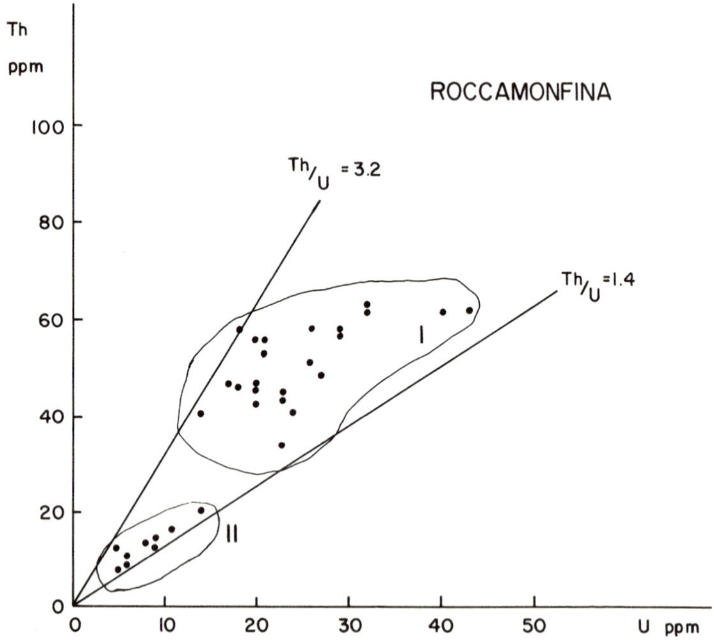

Fig.83. Th–U plot of lavas from Roccamonfina. *I* = leucite bearing rocks; *II* = Trachytes to basalts. (After CIVETTA et al., 1966.)

Fig.83 shows a sharp division into the two groups, supporting the assumption that the two suites derive from different magmas. During the first cycle of activity, U and Th have a behavior opposite to that at Somma–Vesuvio. They in fact increase from leucitites to phonolitic leucite tephrites, also suggesting that different processes were acting at the two volcanoes. Th (8–10 p.p.m.) and U (4–

6 p.p.m.) contents of olivine basalts of the second cycle are contrasting with a mantle origin, rather suggesting that these basalts are extreme mafic differentiates of the original trachytic magma.

High Th and U values are not exclusive of the Campanian area in Italy, because conformable values have been also reported for the Rome area (LUONGO and RAPOLLA, 1965; LOCARDI and SIRCANA, 1967), for the Sicilian volcanoes (CHEMINEE and NORDEMANN, 1963) and Vulture (BONASIA and GASPARINI, 1965).

Mapping of petrologic units

A comparison of field radioactivity data with the geology of the surveyed area often shows that radiometric data are closely related to petrologic units or to particular geologic features. This dependence is locally so close that geologic units can actually be mapped upon radiometric data.

One point must always be kept in mind while interpreting field radioactivity data: more than 90% of the measured gamma flux, when corrected for cosmic rays and atmospheric radioactivity contributions, comes from the upper 10–20 cm of rocks. The contributing layer is of course much more reduced when β^- or alpha particles are considered. This leads to the consequence that radiometric field data are very sensitive to the occurrence of very thin covers of extraneous material. In some cases, a few tens of centimeters of aeolian cover uniformly distributed over different lithologic units have completely masked large differences of radioactivity.

MOXHAM (1963) reports a good example of the correspondence between the natural gamma radiation patterns in Washington County, Maryland, and the lithology of the area. The surveyed area mostly consists (Fig.84B) of lower Paleozoic sedimentary rocks (dolomites, limestones, shales, sandstones, etc.), forming a syncline. The Waverton quartzite and the Catoctin metabasalts are the oldest rocks. They outcrop in stripes along the eastern border of the area. The Paleozoic series is locally covered by unconsolidated Quaternary deposits. Fig.84a shows the radiometric map. It is based upon an airborne survey. The basic detector consisted of six scintillation crystals. The analyzer was gated to give a response to a minimum energy of 0.05 MeV and ungated at the high energy end of the spectrum. Radiations were not discriminated upon their energy. The survey was carried out at a nominal terrain clearance of 167 m along 44 lines oriented north-northeast and spaced at nominal 800 m intervals. The radiometric map clearly shows the same north-northeast trends as the geology of the zone. The highs (>600 counts/sec) are associated with the outcropping of more argillaceous formations, such as the Harpers and Waynesboro formations and the Martinsburg shale. Peak values >1,000 counts/sec were locally obtained. They are associated with the occurrence of a silt loam covering the argillaceous-calcareous bedrock. It is

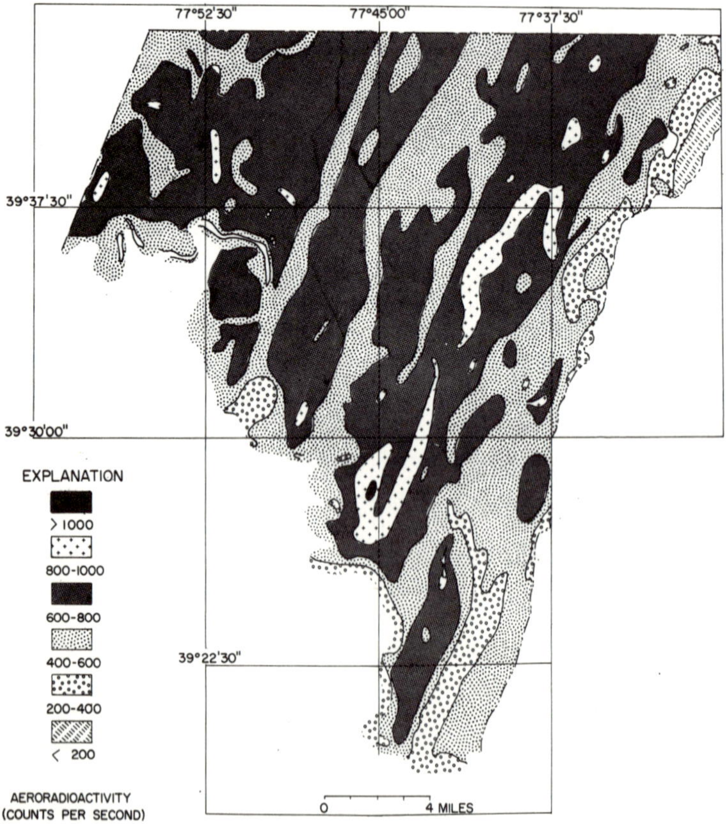

A

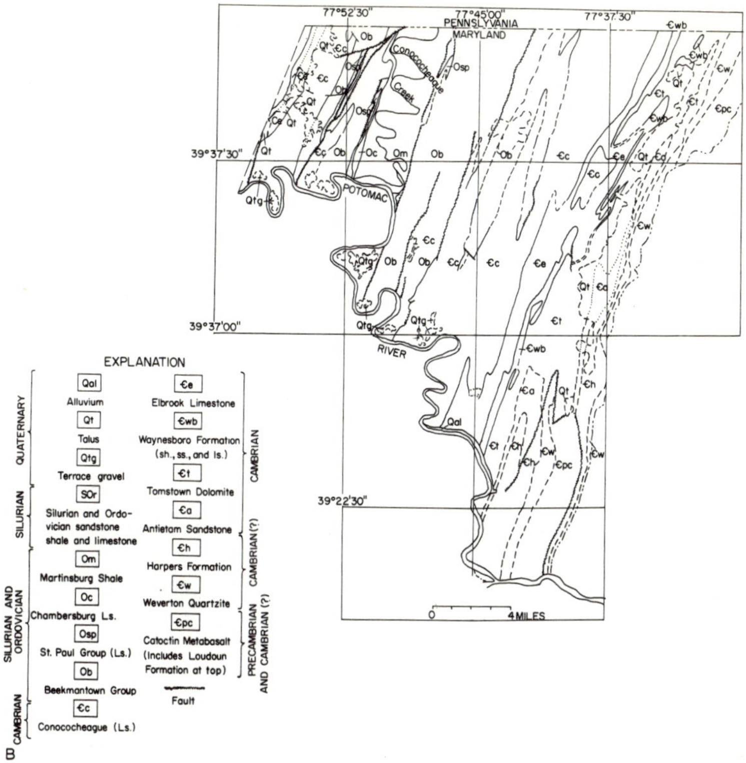

Fig.84. Aeroradiometric (A) and geologic map (B) of the eastern part of Washington County, Maryland. (After MOXHAM, 1963; courtesy of the editor of *Geophysics*.)

probably the result of concentration of K (and eventually also of Th and U in resistate minerals) with the depletion of limestone from the bedrock. Catoctin metabasalts (about 150 counts/sec) and Waverton quartzite (about 300 counts/sec) are the less active Paleozoic rocks. Calcareous formations should exhibit an activity of the same order as basalts. The higher than normal activity of calcareous rocks in this zone is due to the frequent interbedding of argillaceous formations and to the abnormally high K content of limestones. Quaternary unconsolidated deposits have a distribution of activity representative of the main source rocks. The talus in the eastern part of the zone, derived from the rather active

Harpers formation, has locally peak values of 800–1,000 counts/sec. On the other hand, talus of the western section, derived from less radioactive limestones, has comparably lower activities. The strip of low activity along the Potomac river is due to the absorption by water of gamma rays emitted from the stream bed.

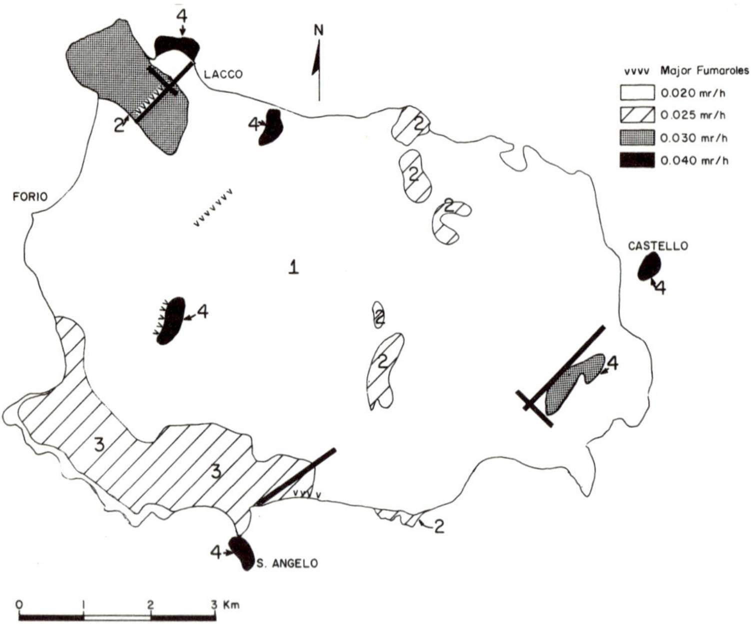

Fig.85. Radiometric map of the island of Ischia. *1* = tuffs, trachytic and latitic lavas, marine debris; *2* = alkali-trachytic lavas; *3* = trachyte and tuff conglomerates; *4* = phonolitic lavas. (After GASPARINI et al., 1961.)

Tectonic features can be outlined by radioactivity surveys under favorable conditions. Faults can be shown when they put in contact units with different activities or radioactive dikes are emplaced along the fault line or Rn ascends through it. The Valley and Ridge physiographic province of Appalachians, Tennessee, contains folded and faulted sedimentary rocks of Paleozoic age. Many of these faults are subparallel and have thrusted the more radioactive argillaceous Rome formation or the Conasauga shale over the less radioactive Knox dolomite. As a result, along the fault a sharp passage from 500–800 counts/sec to 300–500 counts/sec has been observed by an airborne survey. Some of these faults have been traced by radioactivity measurements over more than 50 km (PITKIN et al., 1964).

Fig.85 reports the gamma radiometric map of a volcanic area: the island of Ischia, southern Italy (GASPARINI et al., 1961). It is based upon a field radiometric survey carried out with a GM counter. Measurements were made at the corners of 250 m squares net. Most of the island is characterized by an uniform value of 0.020 mr/h.[1] Lithologically this zone includes the trachytic tuff (green tuff) forming the bulk of the island, volcanic ashes and pumices, latitic and trachybasaltic flows. Alkali-trachytic lavas have activities in the range 0.025–0.030 mr/h and phonolitic lavas in the range 0.030–0.040 mr/h. Alkaline formations are therefore fairly well characterized by their field radioactivity and individual formations can be closely mapped. As pointed out in the former section, the differences observed among different rock units are mainly a consequence of variation of the Th content. U and K change in the same direction as Th, but their variations are comparatively less pronounced. Some narrow zones with abnormally high activities have been observed at the northwest corner (near Lacco Ameno), in the south (near San Angelo), and in the southeast. The characteristic shape of these anomalies along with the close occurrence of fumarolic fields and thermal springs suggest that they are related to fracture lines. The high gamma activity may be a consequence of the ascent of Rn along the fracture. Gamma emitting decay products would be consequently absorbed in the porous surface soil. The decay of activity of Lacco Ameno radioactive thermal waters was found to be consistent with the half-life of ^{222}Rn (IMBO', 1939).

The recent development of portable scintillation spectrometers permits a more detailed differentiation of individual petrologic units on the basis of their K, U, and Th contents (see DOIG, 1968).

---

[1] A roentgen (r) is defined as the amount of X- or gamma radiation producing one electrostatic unit of electricity in 1 cm^3 of dry air at 0°C and standard pressure.

HEAT FLOW STUDIES

Radioactive processes are believed to be the major source of heat within the earth. The dominant contributions are presently due to the natural decay of ^{238}U, ^{235}U, ^{232}Th, and ^{40}K. The heat production is (BIRCH, 1954):

^{238}U series = 0.71 cal./g yr
^{235}U series = 4.3 cal./g yr
^{232}Th series = 0.20 cal./g yr
ordinary U = 0.73 cal./g yr
^{40}K = 0.22 cal/g yr
ordinary K = 27.10^{-6} cal./g yr

These values have been calculated using the mass differences between parent and end products and they have been corrected for the energy carried away by the neutrinos. The energy related to alpha emission accounts for about 90% of the total heat production by the ^{238}U, ^{235}U, and ^{232}Th series. The present heat contributions of ^{238}U, ^{232}Th series, and of ^{40}K in material of terrestrial composition are of comparable magnitude, but ^{40}K, because of its lower half-life, must have had a more dominant role in the past geological history of the earth. Now extinct radioisotopes of appropriate half-life and abundance may have been important early in the geological history of the earth. ^{26}Al (UREY, 1955) is indicated as being most responsible for supplementary radiogenic heat generation in the past history of the earth (for an extensive discussion of extinct or "fossil" radioactivity see KOHMAN, 1956, 1961).

Heat flow determinations are at present unevenly distributed over the earth's surface. The available data show the following features: (*1*) the mean heat flow over the surface of the earth is 1.5 μcal./cm^2 sec; (*2*) average values over continental and oceanic areas are close; (*3*) as to continents, heat flow values are generally lower on Precambrian shields (mean = 0.92 μcal./cm^2 sec) and higher on Mesozoic-Cenozoic orogenic areas (mean = 1.92 μcal./cm^2 sec; (*4*) as to the ocean bottoms, lower average values have been obtained on trenches (0.99 μcal./cm^2 sec), and higher average values on ridges (1.82 μcal./cm^2 sec). For an extensive discussion of the heat flow data and problems the reader is referred to the book edited by LEE (1965) and to the articles by BIRCH (1965) and GIRDLER (1967).

The interpretation of the heat flow data over the surface of the earth requires the knowledge of the distribution of the major radioactive elements within the crust and the upper mantle. Our present knowledge is restricted to outcropping or near surface rocks, so that our ideas on the distribution of radioelements at depth are influenced by what we see at the surface. Furthermore, the data available for the outcropping rocks are not evenly distributed for all rock types; a dominant number of radiometric measurements are on acidic igneous rocks and comparatively few data are available for ultramafic and mafic rocks. Moreover, U concentrations of surfacial rocks may not be representative because of the relatively easy mobilization of U.

The concomitant study of heat flow and radioelements distribution along cores is an approach to the problems of the representativeness of surface samples and of the evaluation of the fraction of total heat flow at any location which can be ascribed to the radioactivity of near surface rocks. F. Birch and co-workers at Harvard University and J. A. S. Adams and co-workers at Rice University are jointly working on a project of this type. Thirteen cores drilled in homogeneous igneous bodies have been analyzed by nondestructive gamma spectrometry for U, Th, and K for a total of 2,365 samples. The maximum depth of the analyzed samples ranges from about 150 to 390 m. The following trends have been observed (ROGERS et al., 1965; ROGERS and ADAMS, 1969b): (*1*) K concentrations are essentially constant throughout each core; (*2*) Th abundances increase downward in four cases, decreases in three cases and are practically constant in all other cases; (*3*) U contents increase downward in four cores, decrease in three cores and are constant in all other cores. No general relationships of the radioelements' abundances with depth is therefore apparent. Evidence of secondary redistribution of U in the surface rock exist for the three cores in the Conway granite, New Hampshire (ROGERS et al., 1965). All the cores have been subjected to runs test in order to find out whether or not any relationship exists between the abundances of any radioelements in adjacent samples. An interdependence has been found for all three radioelements for distances up to 3 m. It has been found that lateral variations in heat flow can be correlated with the radiogenic heat produced within the surface igneous rock formations in which the heat flow was determined. Experimental data fit closely a straight line:

$$Q = a + bH \tag{7.2}$$

where Q is the measured surface heat flow, H is the radiogenic heat produced by the surface rock, a is a constant representing the heat flow when $H = 0$ (i.e., the heat flow from beneath the surface formation), and b is another constant. An interpretation of this relationship is that a surface layer exists where radiogenic heat production varies from place to place. The thickness of this layer (b of eq.(7.2)) is estimated to be 7–11 km in continents. Local variations of heat flow are ascribed to this surface layer, and heat contributions from lower crust and mantle are uniform over wide areas (Roy et al., 1968).

A similar project has been conducted on the Sierra Nevada batholith, California, by the Lawrence Radiation Laboratory, University of California at Berkeley, and by the United States Geological Survey (Lachenbruch et al., 1966; Wollenberg and Smith, 1968; Lachenbruch, 1968). The study included geological mapping, gamma radiometric field survey, laboratory gamma spectrometry on selected samples chosen on the basis of the field data, heat flow and radioactivity determinations along four cores (down to a depth of 500 m), and fission tracks investigations in order to determine the exact radioactive sites. Again in this case no sensitive variations of the radioelements contents with the depth were observed. The values of surface samples averaged within a 1 km radius around the hole are close to the average values of the respective core. Surface heat flow determinations and determinations of U, Th, and K concentrations in surface rocks and boreholes have been made in Western Australia by Hyndman et al. (1968).

A comparison of radioelement abundances in surface and deep samples from Permian sandstones has been carried out by Sikosek and Gojkovic (1966). U contents of marine sandstones are 25% lower in surfacial than in deep samples. On the contrary, U values vary within 5% in continental sandstones. The different behavior is probably a consequence of different sites of the U: in continental sandstones U is fixed mainly inside resistant mineral grains, whereas in marine sandstones a great part can be located in the cement and is therefore easily leachable.

In the models for the interpretation of heat flow data it is often assumed that the concentrations of U, Th, and K in the lower crust are the same as those observed in surfacial acidic rocks. However, even when it may be reasonable to assume a constancy of the bulk chemical composition down to the Mohorovicic level (e.g., in cratons or shield areas), it is likely that the deepest crustal layers are depleted at least in Th and U. When a normal thermal gradient and a crustal depth of 35 km is assumed, the temperature and pressure at that depth will be close to those normally assumed for the existence of the granulite-amphibolite facies of metamorphism. HEIER and ADAMS (1965) have analyzed by gamma-ray spectrometry granulites and amphibolites from Norway for U, Th, and K. They have found that high granulite facies rocks have a depletion of Th and U as compared to the respective low grade metamorphic rocks. LAMBERT and HEIER (1967) have carried out a similar study on rocks from Western Australia. They have observed a significant depletion of U and Th concentrations in felsic pyroxene granulite subfacies rocks. No variation of Th and U abundances with the metamorphic grade has been observed in mafic rocks. In conclusion, mobilization of U and Th by metamorphic processes, suggests that the upper crust should be enriched in U and Th in respect to the lower crust. At least in some parts of the world, seismic evidence exists of the occurrence of a basaltic-type layer in the deep crust. The abundances of radioelements observed in outcropping mafic rocks suggests also in these zones a depletion of Th, U, and K in the deep crust.

We have no direct knowledge of the material forming the upper mantle. Models have been built based on seismic and heat flow data and on the stability and melting relationships of silicate minerals at the pressure and temperature that are believed to prevail at upper mantle levels. Many authors now believe that a peridotitic–dunitic upper mantle is presently that best fitting the available data. Seismic evidence exists that oceanic tholeiitic magmas are generated within the upper mantle (EATON and MURATA, 1960). The upper mantle material must therefore be able to generate tholeiitic basalts. High pressure experiments have shown that peridotitic materials can generate basalts by partial melting with the consequent formation of refractory dunite (see the review by GREEN, 1968). Lateral compositional variations in the upper mantle are suggested by some seismic data.

The surface sampling of upper mantle material is a most complex problem. The following rocks are generally believed to be formed in the upper mantle environment or by magmas generated within the upper mantle and unaffected by crustal contamination: (*1*) oceanic tholeiitic basalts; (*2*) peridotite (lherzolite) nodules in alkali basalts. Their mineralogy indicates a stability at about 10–20 kbar of pressure (see HARRIS, 1967); (*3*) ultramafic inclusions in kimberlite pipes. Their mineralogy indicates a depth of formation of about 150 km (see HARRIS, 1967).

TABLE XXXII

U, Th AND K CONCENTRATIONS IN ROCKS OF PRESUMED MANTLE ORIGIN
(After ROGERS and ADAMS, 1969a,b)

Rock	U (*p.p.m.*)	Th (*p.p.m.*)	K
Oceanic tholeiitic basalts	0.10–0.18	0.18–0.95	0.1–0.7%
Dunite nodules	0.003–0.009	0.05	
Peridotite (lherzolite) inclusions[1]	0.003–0.114	0.014–0.457	12–110 p.p.m.
Pipe eclogites	0.052–0.26	0.22–0.44	
Serpentinites (Mayaguez)[2]	0.014–0.020	0.015–0.021	0.0043–0.0047%

[1] After GREEN et al. (1968).
[2] After ADAMS (1963).

Table XXXII reports the U, Th, and K contents of all these rocks. Tholeiitic basalts are those having a higher radioelement content; this is consistent with the hypothesis of their origin by partial melting of peridotites. In that case, in fact, the "incompatible" U and Th would be concentrated in the liquid phase rather than in the refractory dunite. Because of the low abundances, the analyses are preferentially carried out by neutron activation techniques. The last row reports the abundances determined by gamma-ray spectrometry in serpentinites reached by a drill at Mayaguez,

Puerto Rico (ADAMS, 1963). Some authors suggest that serpentinite may represent residual fractions of the mantle that have subsequently been subjected to hydration to form serpentinite.

Some authors have made calculations of the thermal history of the earth upon the assumption that the earth has a chondritic composition (MACDONALD, 1959, 1961, 1963, 1965; CLARK, 1961). The average concentrations used for these calculations are:

$$U = 1.1 \cdot 10^{-8}\ g/g \qquad Th = 4 \cdot 10^{-8}\ g/g \qquad K = 8 \cdot 10^{-4}\ g/g$$

A remarkable coincidence exists between the present average heat flow and that calculated upon the chondritic model (1.42 cal./cm^2 sec). The validity of the chondritic model has been questioned by GAST (1960) and by WASSERBURG et al. (1964). GAST (1960) noted that the chondritic K/U ratio ($8 \cdot 10^4$) is abnormally higher than the terrestrial value (about $1 \cdot 10^4$). Hence, the chondritic model requires a strong concentration of U with respect to K in the crust and a consequently higher K/U ratio in the mantle than in the crust. This is not supported by the available evidence. The main difference of the chondritic from the terrestrial ratio consists of the relative contribution of ^{40}K. In fact, at present ^{40}K generates about 59% of the total radiogenic heat in chondrites, but only 15.8% in material with terrestrial K/U ratio. This involves also a different thermal history of the earth. The lower half life of ^{40}K in respect of ^{238}U and ^{232}Th causes the ratio of the rates of total heat production 4.5 aeons (10^9 years) ago to the present to be 8.2 for the chrondritic model and 4.5 for the terrestrial model (WASSERBURG et al., 1964).

DISEQUILIBRIA IN NATURAL RADIOACTIVE SERIES

As already discussed in Chapter 4, many cases of radioactive disequilibrium have been observed in nature. The primary cause of most of them is the occurrence, in each series, of radioisotopes of very different chemical properties. Most of the reported cases concern the ^{238}U series. This is a consequence of the presence of the long lived isotopes ^{230}Th and ^{226}Ra ($T_{\frac{1}{2}} = 8.00 \cdot 10^4$ yr and 1,602 yr respectively). The main cause of U–Th fractionation in nature is the higher oxidability of tetravalent U as compared to Th. The uranyl ion, UO_2^{2+}, is more soluble and generally has a greater mobility. General chemical differences (e.g., valences, ionic radii, solubility, etc.) exist between Ra (essentially an alkaline earth) and the Th–U group and these differences play an important role in the geochemical differentiation of Ra. Because of their long half-lives even an episodic removal of any of these isotopes produces long range disturbances of the equilibrium conditions in the ^{238}U series. At least one case of disequilibrium observed in the ^{238}U series cannot be solely ascribed to chemical differences, i.e., the depletion of ^{234}U observed in weathered rocks and soils and the consequent enrichment in fresh and oceanic waters. The causes of this isotopic fractionation are not fully understood. An isotope of Th (^{234}Th) occurs between ^{238}U and ^{234}U; thus the recoil the ^{234}U nuclei has undergone after one alpha emission may be a cause for the differential behavior of the two isotopes.

A minor number of disequilibrium cases have been reported for the ^{232}Th series. This is of course a consequence of the short half-lives of all the ^{232}Th daughters which minimize the effect of any episodic removal.

Deviations of the ^{231}Pa/^{235}U ratio from the equilibrium value have been observed in marine fossils.

A further important cause of disequilibrium is the presence of isotopes of Rn, a gaseous element, in all three series. The chances of a significant Rn escape are highest in the ^{238}U series, because of the higher half life of ^{222}Rn (3.82 days) as compared to ^{220}Rn (55.3 sec) and to ^{219}Rn (4.00 sec). Furthermore, ^{222}Rn has a relatively long lived daughter, ^{210}Pb ($T_{\frac{1}{2}} = 22$ yr), which accumulates in the aereosol and has been found unsupported in snows and glaciers. The determination of equilibrium conditions in natural systems can be useful for many purposes, including absolute dating, sedimentation rate evaluation, estimate of weathering processes, etc. The applicability of gamma-ray spectrometry to equilibrium determinations is restricted by the limited number of sufficiently intense gamma emitters in all three natural series.

As to ^{238}U, it can be determined upon the 0.09 MeV gamma from ^{234}Th. The short time needed for the establishment of equilibrium (about 170 days) warrants a satisfactory reliability in most cases. All the long lived intermediate members except ^{226}Ra are not gamma emitters. ^{226}Ra emits only a 0.186 MeV gamma with 4% absolute intensity and therefore its activity could be determined directly with the relative full energy peak. In any case, the counting rate (or at least the full energy counting rate) under the peak is usually low and furthermore, under equilibrium conditions, about 36% is due to 0.185 MeV gammas from ^{235}U. It is therefore much more convenient to use the more prominent part of the spectrum due to post-^{222}Rn gamma emitters. Equilibrium down to ^{214}Bi is established within about 32 days and, therefore, if a continuous escape of ^{222}Rn can be ruled out, a more accurate determination can be carried out in this way. ^{234}U and ^{230}Th cannot of course be determined directly by gamma spectrometry, but only upon their gamma emitting daughters with which equilibrium conditions must be known or assumed. Alpha spectrometry is the only technique allowing a direct determination of both isotopes and of ^{238}U. ^{210}Pb is a gamma emitter, but the low energy (0.046 MeV) and absolute intensity (4.1%) are a serious drawback for its determination by gamma-ray spectrometry.

For the ^{232}Th series the main problem is just the determination of the progenitor. The two first members of the series (^{232}Th and ^{228}Ra) are in fact the longest lived and have practically no gamma emission. The third member, ^{228}Ac, has a sufficiently intense gamma emission. The short time needed for the establishment of equilibrium throughout the series limits however the chances of significant disturbances of equilibrium conditions to the case of a practically continuous removal of any of the members. Equilibrium between ^{232}Th and ^{228}Th, or between pre- and post-^{220}Rn members, can be checked upon the gamma emissions from ^{228}Ac and from the post-^{220}Rn gamma emitters.

The two longest lived members of the ^{235}U series (^{235}U and ^{231}Pa) are both gamma emitters. But in this case it is the low natural isotopic ratio of ^{235}U which inhibits a direct determination, because the contributions from the members of this series to natural gamma spectra are largely overshadowed by ^{232}Th and ^{238}U gamma emitting daughters and by ^{40}K.

YOKOYAMA et al. (1968) have determined the oceanic sedimentation rate by the measurement of the ^{230}Th/^{232}Th ratio along a core from the Indian Ocean. This method, using gamma spectrometry, is based on the fact that U is more soluble than Th in sea waters, where it forms complex carbonate ions, whereas Th preferentially precipitates on the bottom where it is absorbed in sediments. Unsupported ^{230}Th decreases with the time according to its own decay rate so that, if the sedimentation rate has not changed appreciably during all the time, the decrease of the ^{230}Th content with the depth is a measure of the sedimentation rate. The assumption of a constant sedimentation rate for Th is checked by determining the ^{232}Th all along the core. Non-destructive gamma-ray spectrometry has been carried out by a NaI(Tl) spectrometer on about 5 cm high pieces of cores. ^{232}Th has been determined by the 0.24 MeV gammas from ^{212}Pb, and ^{230}Th by gammas from ^{214}Bi and ^{214}Pb. Equilibrium conditions from ^{232}Th to ^{212}Pb and from ^{230}Th to ^{214}Bi have therefore been assumed. In order to allow for the physical establishment of equilibrium between ^{230}Th and ^{226}Ra, the upper 1.7 cm of the core, which have been estimated to correspond to a sedimentation period of about 10^4 yr, have been rejected. The ^{232}Th has been found reasonably constant throughout the core, indicating therefore an essentially constant sedimentation rate for Th. ^{230}Th content (and of course the ^{230}Th/^{232}Th ratio) decreases with the depth down to a limit value corresponding to the equilibrium with the ^{238}U present in the sediments. The supported fraction has been estimated to be about 5% of the total ^{230}Th in surfacial sediments. A sedimentation rate of 1.7 mm/1,000 yr has been inferred, which is consistent with the values obtained by other techniques.

Another example concerns the study of the ^{238}U–^{226}Ra relationship in the lavas from Vesuvius. An average decrease of the radioactivity with increasing age of lavas from Vesuvius was observed in total radioactivity measurements (JOLY, 1909; IMBO', 1955). Nuclear emulsion measurements have then shown that this was essentially due to a fairly regular decrease of the ^{226}Ra content (CASERTANO, 1952; FERRETTI SFORZINI et al., 1955). It was therefore hypothesized that all the pre-^{226}Ra members of the series were missing in Vesuvian lavas. Gamma spectrometry and mass spectrometry on lavas erupted in historical times (VITTOZZI and GASPARINI, 1964; RAPOLLA and VITTOZZI, 1968; OVERSBY and GAST,

1968) have shown the presence of ^{238}U. The $^{226}Ra/^{238}U$ ratio, however, exceeds the equilibrium value. When the logarithm of the excess ^{226}Ra (evaluated upon the 0.61 MeV full energy peak) is plotted against the age of the respective lavas, the experimental points fit closely a straight line. The experimental decay constant is of the order of 10^{-11} sec^{-1}, i.e., of the same order as that of ^{226}Ra ($1.37 \cdot 10^{-11}$ sec^{-1}), indicating that: (*1*) the unsupported member of the series is actually ^{226}Ra, and therefore a fraction of all the members from ^{238}U to ^{226}Ra is actually missing; (*2*) either the removal of ^{238}U along with its missing daughters occurred practically at the same moment of the history of each lava or ^{226}Ra is supplied to the magma with a constant rate so that an equilibrium condition is reached and no variation of ^{226}Ra within the magma occurs.

The causes of this disequilibrium are not understood as yet. Any removal of U isotopes could be explained by oxidation and mobilization in the gas phase. The removal of ^{230}Th is not easily justified on the same basis, although its behavior in gaseous transfer phenomena indicates a certain tendency to go along with the gaseous phases.

LUNAR GAMMA SPECTROMETRY

The absence of an atmosphere dense enough to absorb significantly average energy gamma rays makes possible the analysis of the gamma emission from the lunar surfaces directly from an orbiting satellite. Information that can be derived from the knowledge of the gamma spectrum of the surface of the moon include: (*1*) the approximate chemical nature of the surface rocks as shown by the gamma emission intensity due to the natural noncosmogenic radioisotopes; (*2*) presence of possible radioactive surfacial heterogeneities; (*3*) concentrations of some major rock forming elements upon characteristic gamma emissions from cosmic rays induced reactions and from cosmogenic isotopes. Compared to the earth, where the atmosphere protects surface rocks against primary cosmic radiation, the moon surface is directly exposed to the bombardment from the primary component of both cosmic and solar origin. As a consequence, gamma rays are emitted mostly from these induced reactions and from a number of cosmogenic radioactive nuclides, such as those observed in meteorites. The only difference is that the bombardment is practically continuous even

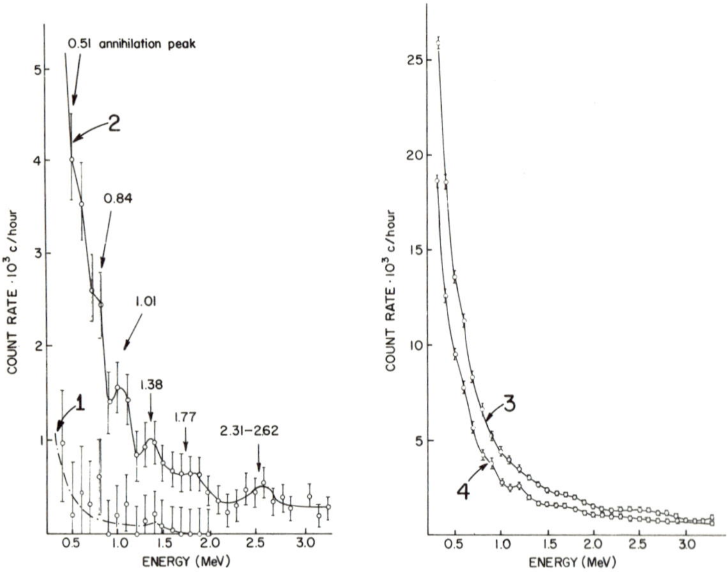

Fig.86. Gamma-ray spectrum recorded by the cosmic station "Luna 10." *1* = gamma spectrum ascribable to emissions from non-cosmogenic nuclides in the rocks of the lunar surface; *2* = gamma spectrum of the lunar surface obtained upon subtraction of curve *4* from curve *3*; *3* = gamma spectrum of a zone of the invisible part of the moon; *4* = background spectrum produced by cosmic rays interaction with the material of the satellite, recorded at 230,000 km from the earth and corrected for the screening effect of the moon. (After VINOGRADOV et al., 1966.)

during the measurement; both short and long lived radionuclides are present, their concentrations being proportional to the abundance of the respective target nuclei and to the effective cross sections.

The first measurement of the gamma spectrum of the lunar surface has been carried out from the Russian cosmic station "Luna 10", between April 5 and 21, 1966. The results have been discussed by VINOGRADOV et al. (1966). The instrumentation consisted basically of a 32 channel NaI(Tl) spectrometer. The detecting system was formed by a 30 mm diameter by 40 mm high cylindrical NaI(Tl) crystal enclosed within a thin plastic scintillator. The purpose of this arrangement was to reduce the effect of charged particles. Gamma rays are in fact essentially unaffected by the thin plastic layer, whereas charged particles produce coincident effects in both detectors and can be therefore eliminated by anticoincidence circuitry. The 32-channel pulse height analyzer was pre-set so that it could inspect a wider (from 0.3 to 3.1 MeV) or a smaller (from 0.15 to 1.5 MeV) energy range. It could be switched from one position to the other by radio commands from the earth.

The cosmic station recorded six gamma spectra in the energy range 0.3–3.1 MeV on different areas, covering both "continents" and "seas" in the visible and back side of the moon. A gamma spectrum was recorded at about 230,000 km from the earth (about 70,000 km from the moon). This spectrum is practically all formed by gamma rays following cosmic interactions with the material of the cosmic station. A very minor part of it is to be ascribed to natural gamma emitting impurities in the spacecraft.

Curve *3* of Fig.86 shows the spectrum obtained on April 8, while overflying an area of the back side of the moon. The spectrum consists of the following contributions: (*1*) background due to cosmic-ray induced and natural radiations from the cosmic station itself; (*2*) radiations from the surface of the moon following cosmic ray interactions; (these two components can be split into two parts: characteristic gamma rays from nuclear reactions and from cosmogenic radionuclides; and secondly, a continuum distribution due to muon decay, electron emission, etc.); (*3*) radiations from natural non-cosmogenic radioisotopes, i.e., practically from members of the ^{238}U, ^{235}U, and ^{232}Th series and from ^{40}K.

Curve *4* is the spectrum obtained at about 230,000 km from the earth, corrected for the screening effect of the moon. Curve *2* is the spectrum obtained upon subtraction of curve *4* from curve *3*. It represents the net contribution from the lunar surface alone. The main difference from the earth's surface gamma spectrum is the occurrence of a remarkably higher than 1.5 MeV component. The higher than 1.5 MeV part of the spectrum amounts to about 20% for the lunar surface spectrum compared to about 5% for the earth surface spectrum when measured with the same instrumentation. Many full energy peaks are distinguishable. They are believed to be derived from gamma emissions after cosmic-ray induced reactions with O, Si, Mg, Al. The main contributing cosmogenic radio-nuclides are expected to be ^{14}O, ^{19}O, ^{20}F, ^{22}Na, ^{24}Na. A rough estimate of the gamma dose on the lunar surface indicates that it should be about 1.5–2 times higher than that on average terrestrial granitic rocks.

In order to evaluate the contribution from non-cosmogenic natural isotopes, all the cosmic ray induced components must be subtracted. Since this spectrum is experimentally unknown, it has been assumed that it has the same shape as the cosmic ray induced spectrum on the spacecraft. This was in fact built of light alloys, containing as major components Si, Al, and Mg, i.e., approximately the same major rock forming elements. The background spectrum recorded along the earth–moon trip was then fitted to the energy region above 2 MeV (where the natural non-cosmogenic contribution can be considered for this case to be negligible) and then subtracted from curve *2*. Curve *1* is the resulting spectrum, representing less than 10% of the total gamma intensity. Calibration with U, Th, and K standards has allowed one to predict the response of the spectrometer while recording in orbit over rocks with different U, Th, and K concentrations. The obtained response is comparable to that predicted for basaltic rocks. The overwhelming influence of the cosmic ray induced component severely restricts the accuracy of the determination of natural non-cosmogenic radioactivity, so that more meaningful values can be obtained only after some lunar samples are analyzed in the laboratory.

No variations higher than 40% have been observed over the surveyed areas, but, also in this case, the cosmic ray induced component could completely mask non-cosmogenic radioactivity variations. The obtained data do allow one to exclude with a relatively high confidence, the presence of granitic type rocks over the surveyed parts of the surface of the moon.

TEKTITES AND METEORITES

Common problems in the measurement of radioactivity in tektites and meteorites are the small amount of sample usually available and the high desirability of nondestructive analyses. Gamma-ray spectrometric techniques are a frequent choice, although the low counting rates severely restrict the accuracy of the determinations. Improved geometry arrangements, very low background instrumentation and coincidence counting are usual expedients in order to improve the sensitivity and the accuracy. Close geometric conditions for standards and samples can be reproduced by mock up arrangements, such as those described by VAN DILLA et al. (1960); BALACEK (1966); TOBAILEM et al. (1967).

The main purpose of the radioactivity measurements on tektites is to determine the U and Th contents in order to give a further requirement for any of the hypotheses about their origin.

The main characteristics of tektites can be summarized as follows:

(*1*) They are small glass fragments (from a few mm to about 10 cm in size) usually with rounded surface features, sometimes with tear drop shapes, and surface sculpturing interpreted as due to hypersonic velocity flights through an atmosphere.

(*2*) They occur in strewn fields, the major ones being in Czechoslovakia (tektites from this field are named moldavites), Australia (australites), southeast Asia (philippinites, indochinites, billitonites), southeastern U.S.A. and Texas (bediasites), and the Ivory Coast, Africa. K–Ar dates are about 0.7 megayears for tektites from Australia and southeast Asia, about 1.2 megayears for Ivory Coast tektites, about 15 megayears for moldavites, and about 34 megayears for bediasites. K–Ar dating of moldavites are in excellent agreement with K–Ar age of the Ries-Nördlingen impact crater in Bavaria, and Ivory Coast tektites age agrees closely with that of Bosumtwi impact crater in Ghana (for a more detailed discussion see SCHAEFFER, 1966).

(*3*) Their chemical composition is close to that of terrestrial sialic rocks.

Many hypotheses have been put forward about the origin of tektites, favoring both terrestrial and extraterrestrial origin. Terrestrial origin is strongly supported by their chemical similarity with sialic crustal rocks. The occurrence in strewn fields and the similarity in K–Ar ages of tektites from Czechoslovakia and Ivory Coast and two neighboring impact craters are additional evidence for their terrestrial origin, suggesting that they could be related with the impact of large meteoritic bodies. Aerodynamic considerations are the major arguments speaking against the terrestrial origin. For a more detailed discussion the reader is referred to the book edited by O'KEEFE (1963).

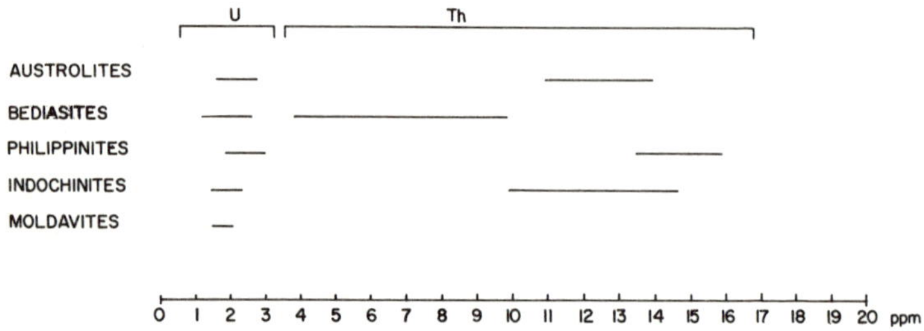

Fig.87. Ranges of U and Th contents of tektites.

U and Th determinations have been mostly made by gamma-ray spectrometry and have been recently summarized by ROGERS and ADAMS (1969a, b) and by BALACEK (1966). Fig.87 reports the U and Th variation range, considering only the most recent gamma spectrometric data (CHERRY and ADAMS, 1963; ROWE et al., 1963; BALACEK, 1966). U and Th abundances range from 1.1 to 3.0 and from 3.9 and 15.9 p.p.m. respectively. They are, therefore, of the same order as terrestrial crustal sialic rocks. The Th/U ratios are generally higher than 4, appearing to be therefore higher than that of average crustal material, but not at all unusual for crustal rocks. The U and Th content of extraterrestrial material is generally lower than those of tektites by a factor of ten or more; the measurements of radioactivity strongly support the hypothesis of a terrestrial origin. The data reported by VINOGRADOV et al. (1966) about the gamma emission from the surface of the moon (see pp.217–220) are not consistent with the hypothesis of a lunar origin.

No definite U variations among the different strewn fields are apparent and the respective ranges are largely overlapping. On the contrary, variations in the Th content seem to be real in indicating higher abundances in philippinites and a slight deficiency in bediasites. The low Th content of bediasites is reflected also in a lower Th/U ratio, whereas philippinites, indochinites, and australites have very similar Th/U ratios. This, along with the close K–Ar ages, would substantiate the idea that they constitute a single strewn field. Autoradiographic investigations (BALACEK, 1966) have shown a uniform distribution of radioactive sites.

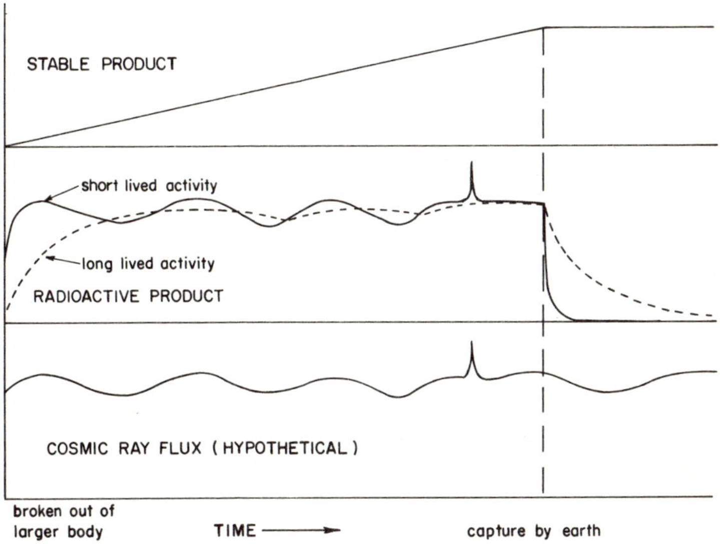

Fig.88. Qualitative variation of the concentrations of cosmogenic nuclides along the history of a meteorite. (After SCHAEFFER, 1962.)

The measurements of radioactivity on meteorites have not only the purpose of the determination of U and Th abundances, but also of determining the concentrations of cosmogenic radionuclides. Fig.88 (from SCHAEFFER, 1962) reports the qualitative variation of stable and radioactive cosmogenic nuclides during the history of a meteorite. From the moment it has been broken out of a larger body, the high energy bombardment by primary cosmic rays begins and consequently cosmogenic nuclides begin to form. The abundances of stable cosmogenic nuclei increase with the time at a rate proportional to the cosmic ray flux on the meteorite, whereas radioactive nuclei tend to an equilibrium concentration practically reached after several half-lives. If the cosmic rays flux on the meteorite changes either because it is not constant with the time or because of variations in different regions of the space, the abundances of shorter lived isotopes will follow more or less promptly the variations of the cosmic ray flux. Longer lived radionuclide abundances will be less affected, tending to smooth and average the changes. From the moment the meteorite is captured by the earth, the concentrations of stable cosmogenic isotopes do not change anymore, whereas those of radioactive isotopes will decrease according to the respective decay constants. Many important data can be obtained from the measurement of the cosmogenic isotopes, including constancy of cosmic rays both in space and time during the past million years, exposure ages (i.e., span of time it has been exposed to primary cosmic radiation), and terrestrial age (i.e., age of capture by the earth) of the meteorite, space erosion, etc.

TOBAILEM et al. (1967) have carried out gamma spectrometric determinations on a 113 kg fragment of the meteorite fallen at St. Severin (Charente, France) on June 27, 1966, starting only 60 hours after its fall. The specific activity of many short lived radionuclides, such as ^{24}Na (15.0 h), ^{52}Mn, ^{48}V, ^{58}Co, ^{46}Sc, ^{54}Mn, along with ^{22}Na, ^{60}Co, ^{26}Al, and ^{40}K, has been determined. Specific activities evaluated for the time of fall range from <6 (^{58}Co) to 85 (± 10) (^{54}Mn) decays/min kg for all the isotopes, but ^{40}K (1,720 $\pm$ 150 decays/min kg). When dealing with meteorites with terrestrial ages of the order of years, ^{26}Al and ^{22}Na are the isotopes more often determined by gamma spectrometric techniques. Both of them decay prevalently by positron emission promptly followed by 1.83 (^{26}Al) and 1.28 (^{22}Na) MeV gamma rays. γ–γ coincidence equipment can therefore be used, one of the detectors triggering the coincidence circuit after the absorption of a 0.511 MeV positron annihilation quanta (ANDERS, 1960).

Fig.89 reports the range of variation of U and Th contents of achondrites, chondrites, and iron meteorites. As they are usually a fraction of p.p.m., gamma spectrometric techniques cannot be

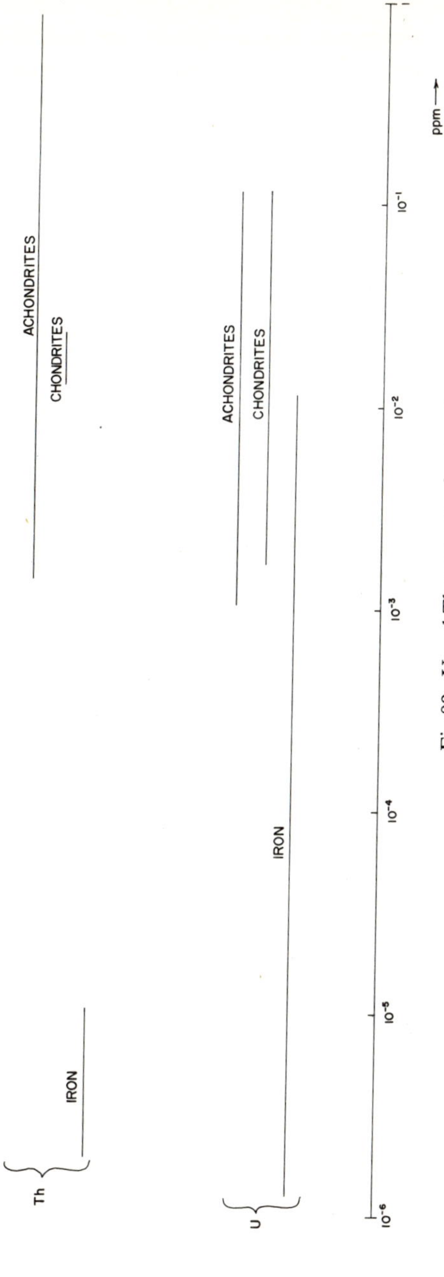

Fig.89. U and Th contents of meteorites.

applied with any confidence, except perhaps for the most active achondrites and chondrites. Most of the determinations have been carried out by neutron activation techniques.

Th/U ratios for stony meteorites are in the range of 3–4. The average ratio is supposed to be representative of all the solar system. Some authors (MacDonald, 1959, 1961, 1963, 1965; Clark, 1961) have assumed that the average abundances of Th, U, and K in chondrites represent the average abundances within the earth.

ECONOMIC CONSIDERATIONS

A great majority of total and spectrometric gamma detectors are in industrial operation at present. Unfortunately very little of this industrial experience with rocks and ores has been made available to the public in any form. The industrial applications include among others: (*1*) evaluation of K and U ores; (*2*) location of U anomalies on a regional and localized basis; (*3*) the use of radioactive minerals, e.g., zircon, as "pathfinders" for deposits of other placer minerals; (*4*) the use of radioactive "key beds", e.g., black shales and bentonites, to interpret subsurface stratigraphic sections and structural relationships for petroleum exploration (see Campbell, 1961).

The evaluation of K ores in the subsurface by gamma-ray spectrometry is easily reduced to practice because such ores as sylvite (KCl) are over 50% K and the K photopeak stands in high contrast to the usually minor amounts of thorium and uranium found in evaporite sequences. Gamma spectrometry is particularly suited to evaluating thin and spotty ore bodies that might be amenable to solution mining. In such cases, subsurface gamma spectrometric logging should be superior to core analysis because the annular volume measured in the hole is considerably larger than the core taken out. Gamma spectrometry is also useful in monitoring any milling or concentrating process.

The evaluation of U ores is complicated by the possible radioactive disequilibria in the ^{238}U series (see Dodd, 1966; Dodd et al., 1967; and Rosholt, 1967). A critical point is whether or not the ore deposit being evaluated is in secular radioactive equilibrium as a whole. If the deposit as a whole is in equilibrium, the average of all gamma logging will correctly evaluate the U in the deposit, and disequilibria that cause overestimation of U will just be balanced by disequilibria that cause underestimation. DODD et al. (1967) have comprehensively reviewed the instrumentation and computer programs for U ore evaluation. CZUBEK and DUMESNIL (1968) among others have suggested a possible direct U ore determination in the borehole on the basis of U absorption in the low energy range of the gamma spectrum.

Ground and aerial gamma spectrometric surveys can be used to localize regional or deposit anomalies (see Chapter 3, pp.88–99). ADAMS (1968) has surveyed the 400 miles (660 km) of Texas beach sands and has suggested a gamma spectrometric–magnetometer combination for the exploration for placer sands. CAMPBELL (1961) has reviewed the applications of radioactivity well logging in petroleum exploration.

It should be noted that the relatively high original cost of gamma spectrometric instrumentation is more than offset by the low cost per analysis and the great advantage of having the results of the analysis in the field at the sampling point. Having the results of the analysis in the field permits one to change the sampling pattern and density as well as immediately test hypotheses about the U mineralization being related to color, texture, veining, structure, etc. It should be further noted that with many gamma spectrometric studies there are so many data that one deals with two levels of statistics: (*1*) the lower level is concerned with the counting statistics at a single station or during a single time interval of a traversing instrument; (*2*) the higher level deals with a population of analyses for an entire deposit, or rock unit, or batholith. It is the higher level of statistics that deals with volumes of rock on an economic scale (ROGERS and ADAMS, 1963).

APPENDIX 1

Absorption Coefficients from the Data Reported by GRODSTEIN (1957)

All the photoelectric absorption cross sections are evaluated for K, L, and M shells. Compton cross sections account for both coherent and incoherent scattering.

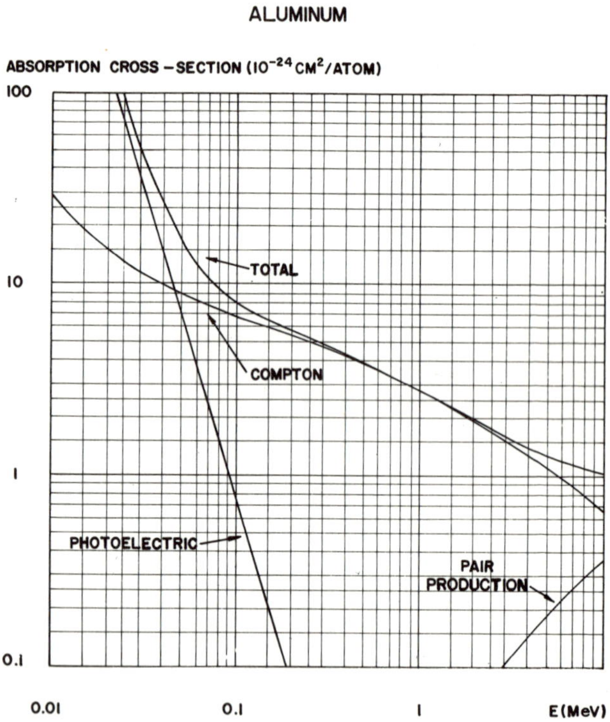

Fig.90. Aluminum ($Z = 13$).

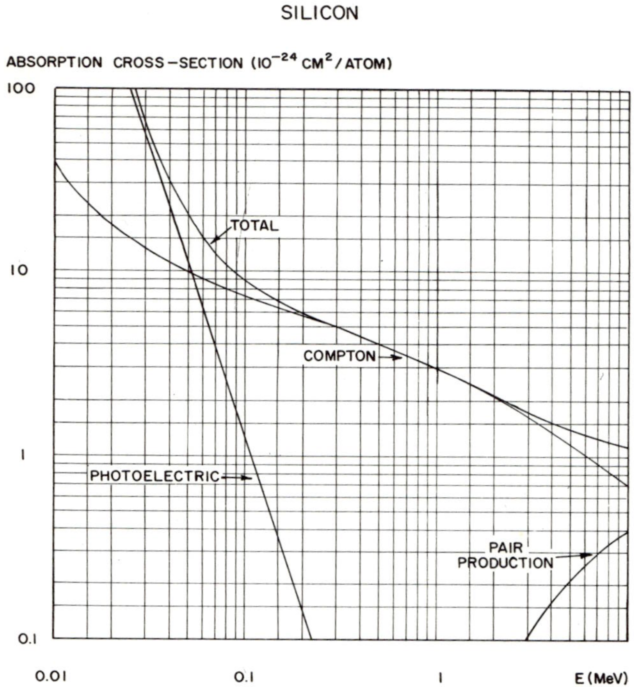

Fig.91. Silicon (Z = 14).

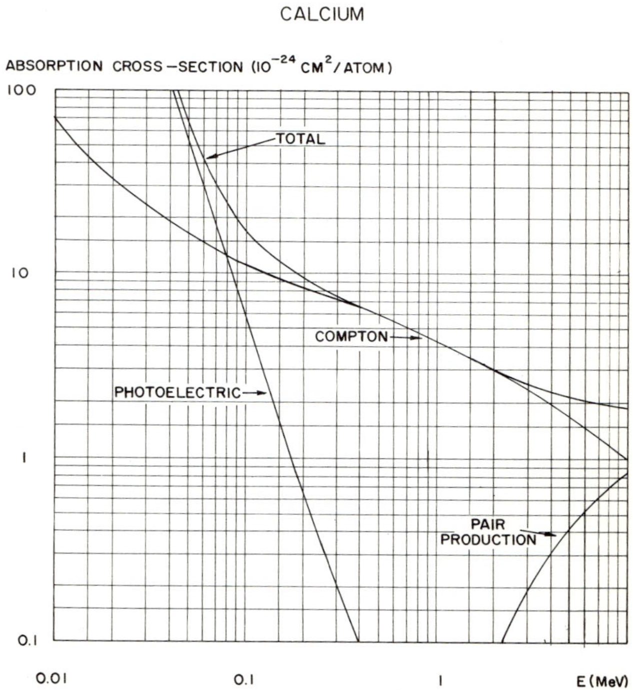

Fig.92. Calcium (Z = 20).

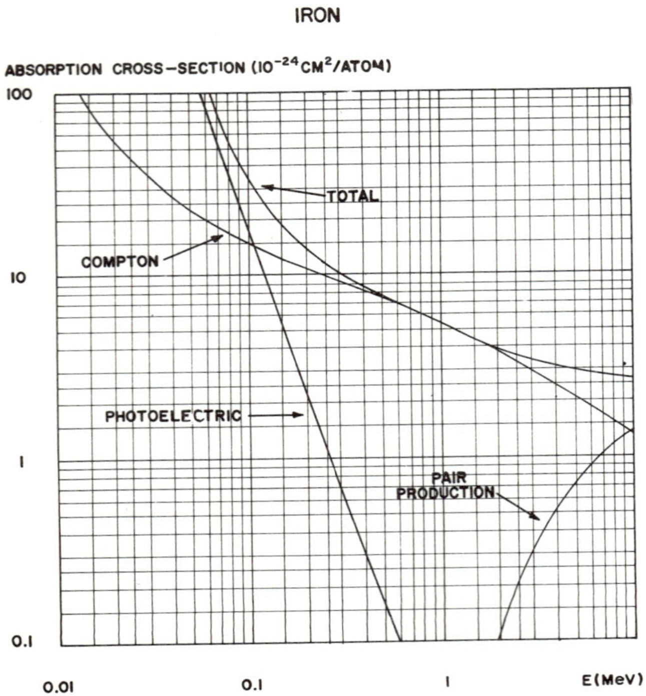

Fig.93. Iron (Z = 26).

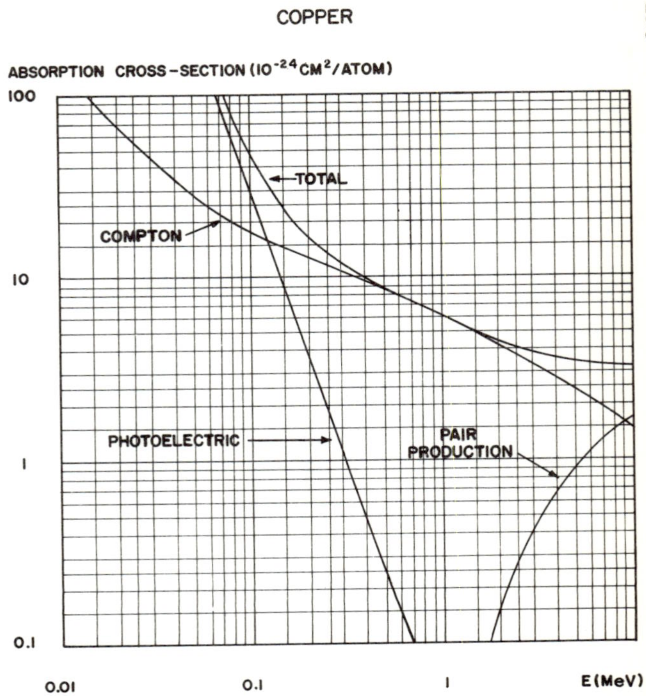

Fig.94. Copper (Z = 29).

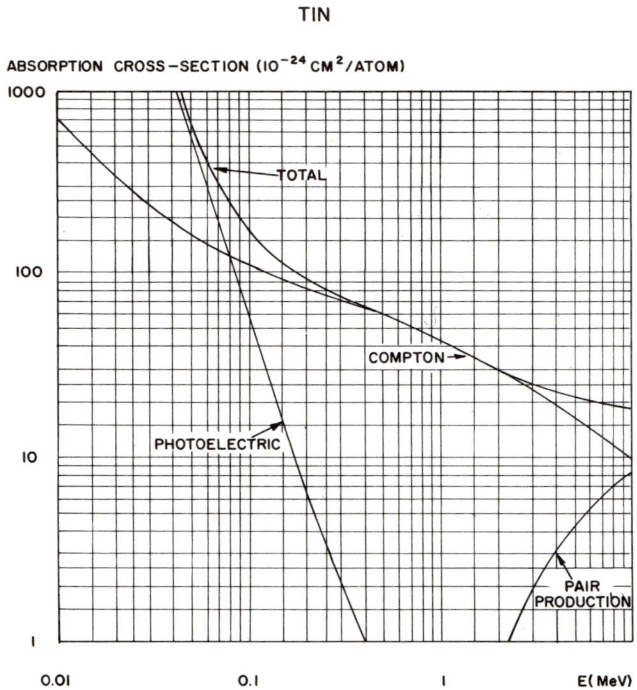

Fig.95. Tin ($Z = 50$).

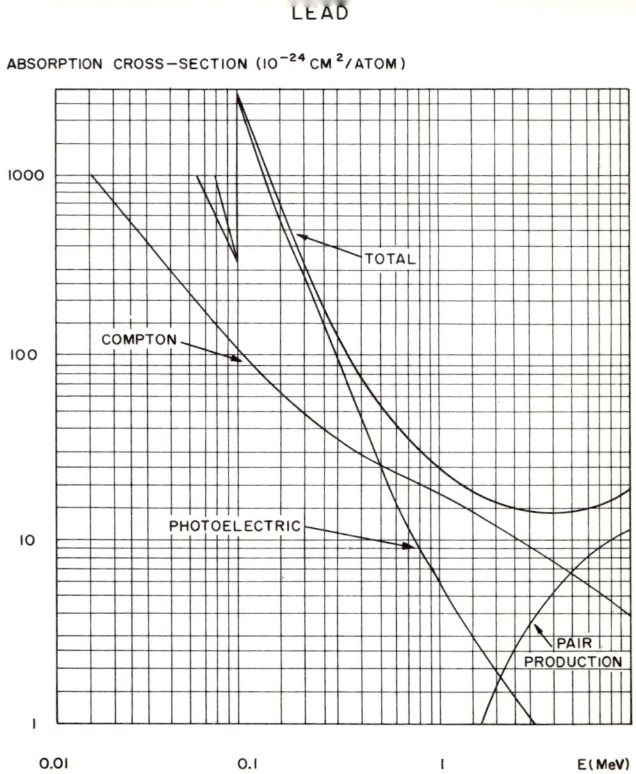

Fig.96. Lead ($Z = 82$).

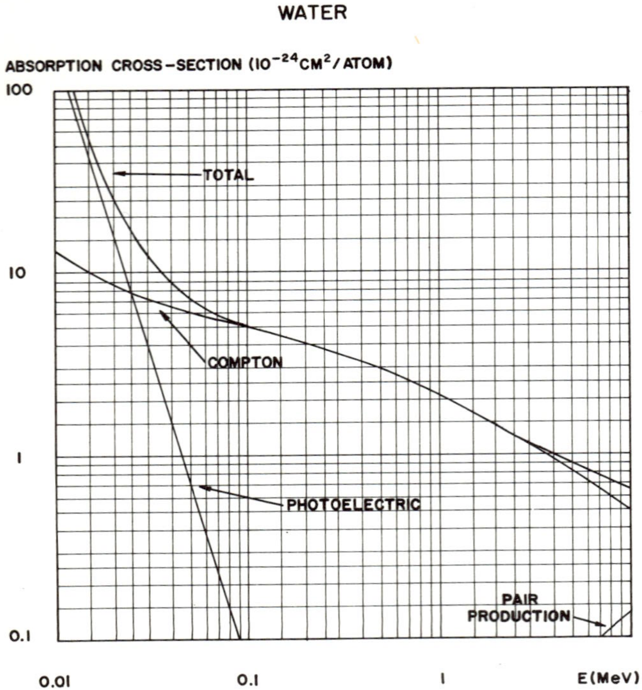

Fig.97. Water

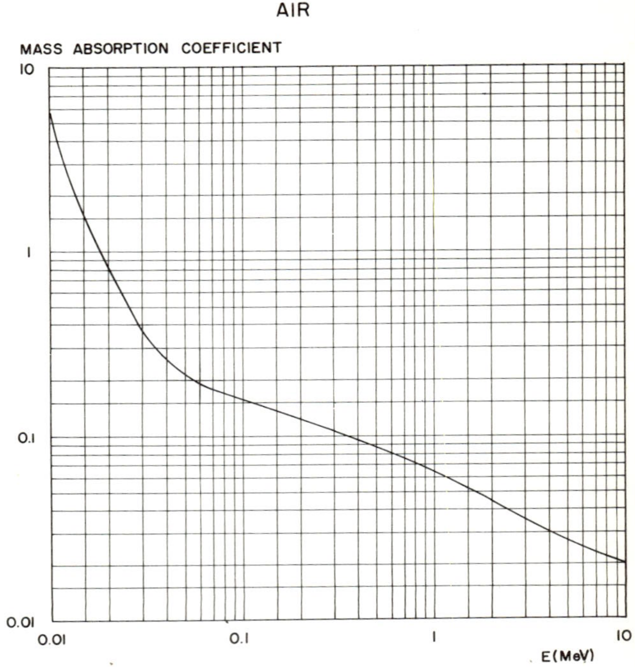

Fig.98. Total mass absorption coefficient of air (N: 77.5%; O: 23.2%; Ar: 1.3% by weight).

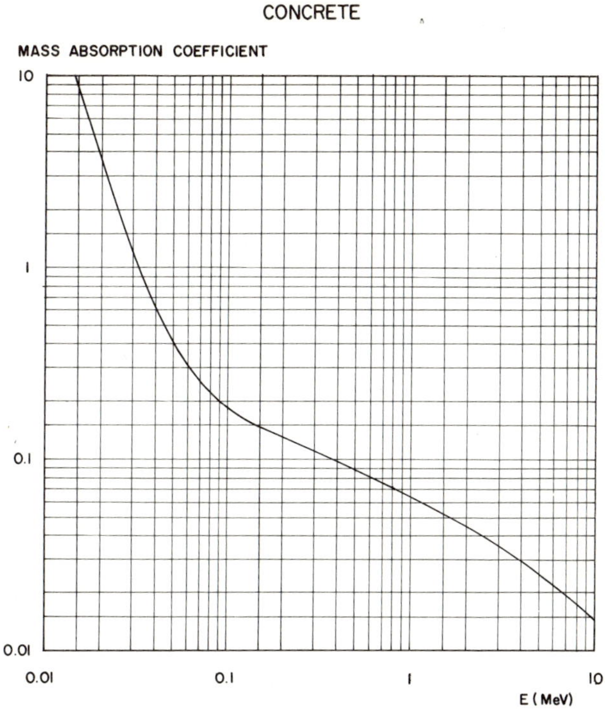

Fig.99. Total mass absorption coefficient of concrete (H: 0.56%; O: 49.56%; Si: 31.35%; Al: 4.56%; Ca: 8.26%; Fe: 1.22%; Mg: 0.24%; Na: 1.71%; K: 1.92%; S: 0.12%) density: 2.35 g/cm^3.

APPENDIX 2

Some Gamma Spectra of Natural Isotopes and Fall-out Fission Products

Samples marked IFTUN have been analyzed at the Istituto di Fisica Terrestre of the Università di Napoli by the instrumentation described in Chapter 3, pp.33–34. A 0.5 mm thick aluminum Marinelli hat container has been used. Each sample has been accommodated in a layer approximately 2 cm thick, surrounding the crystals. U and Th samples were diluted with pure $CaCO_3$ in powder.

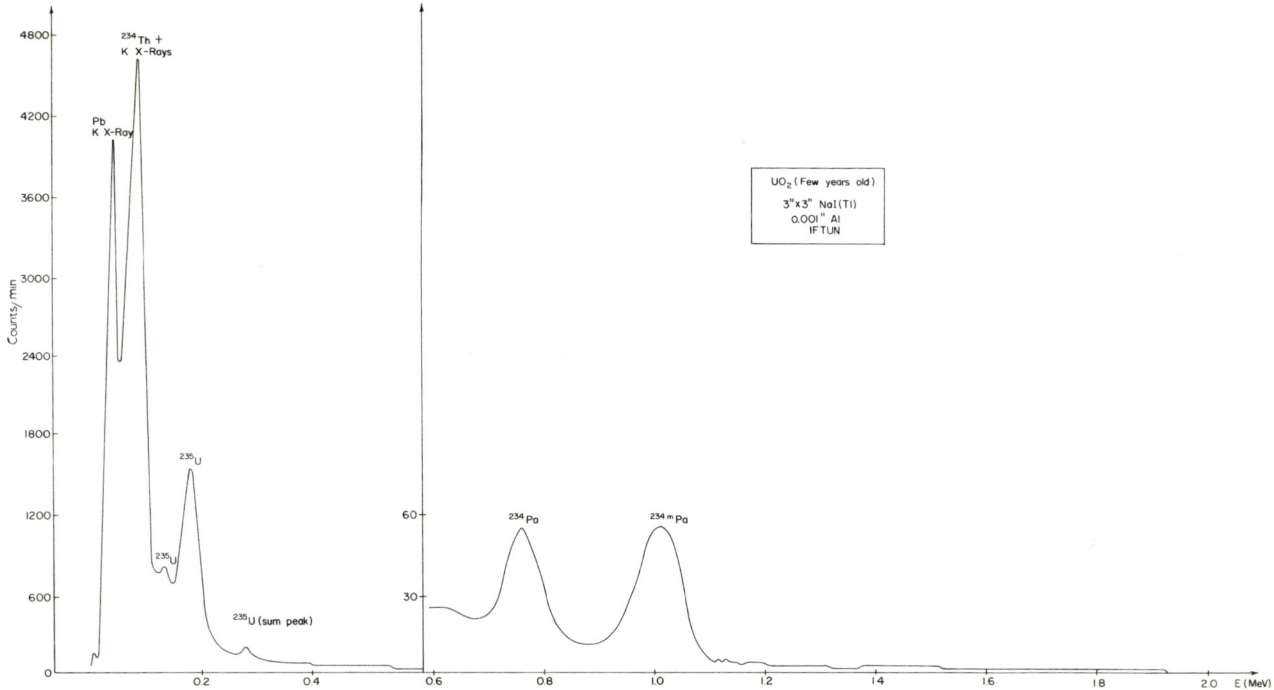

Fig.100. Freshly prepared UO_2, which has not yet grown up appreciable activity from ^{226}Ra and daughters.

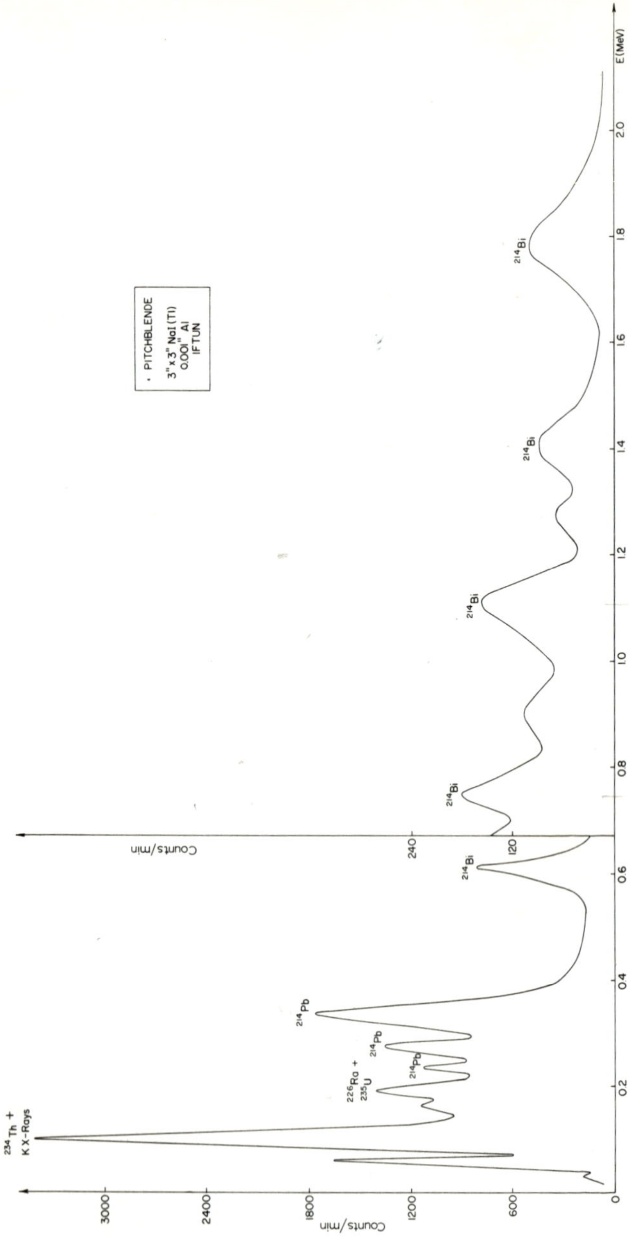

Fig.101. Pitchblende with ^{238}U series in secular equilibrium.

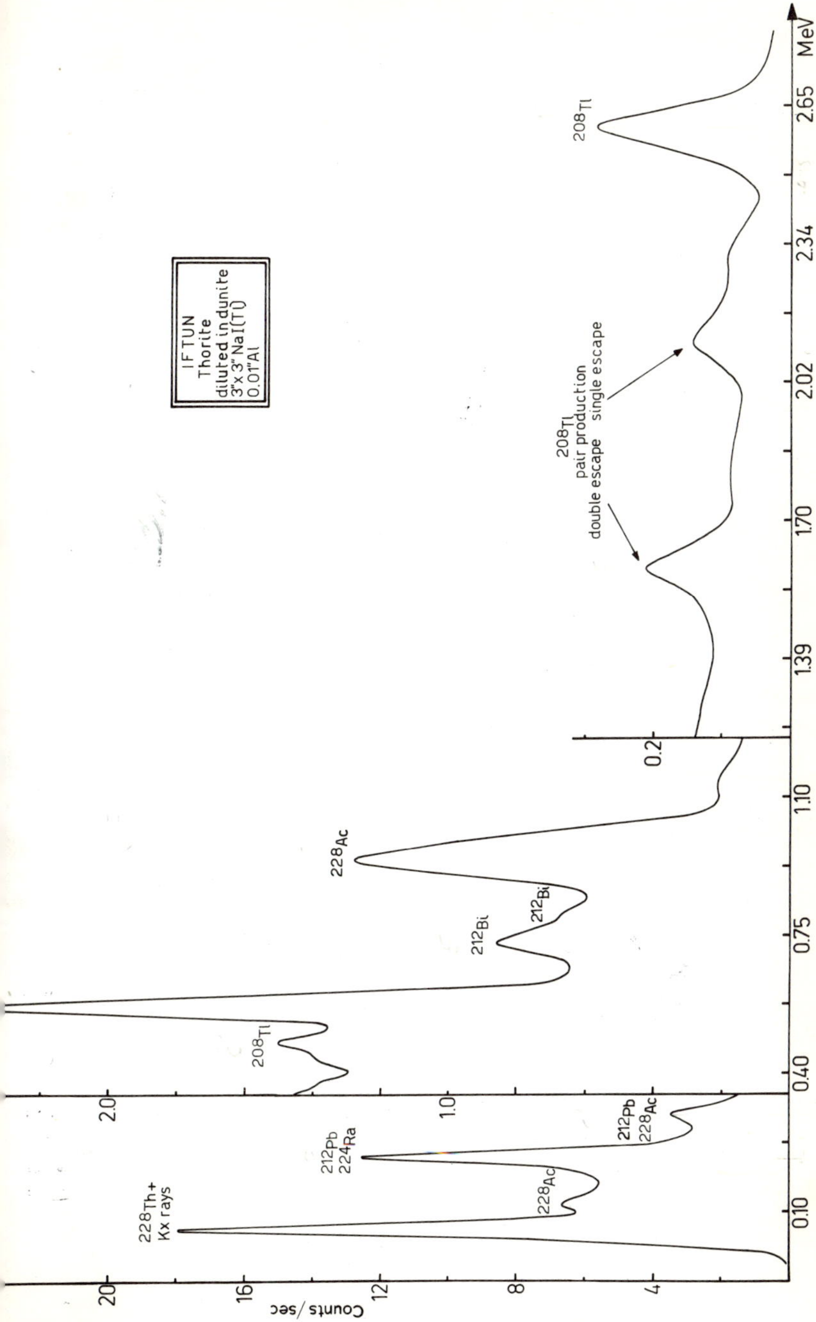

Fig.102. Thorite with ^{232}Th series in secular equilibrium.

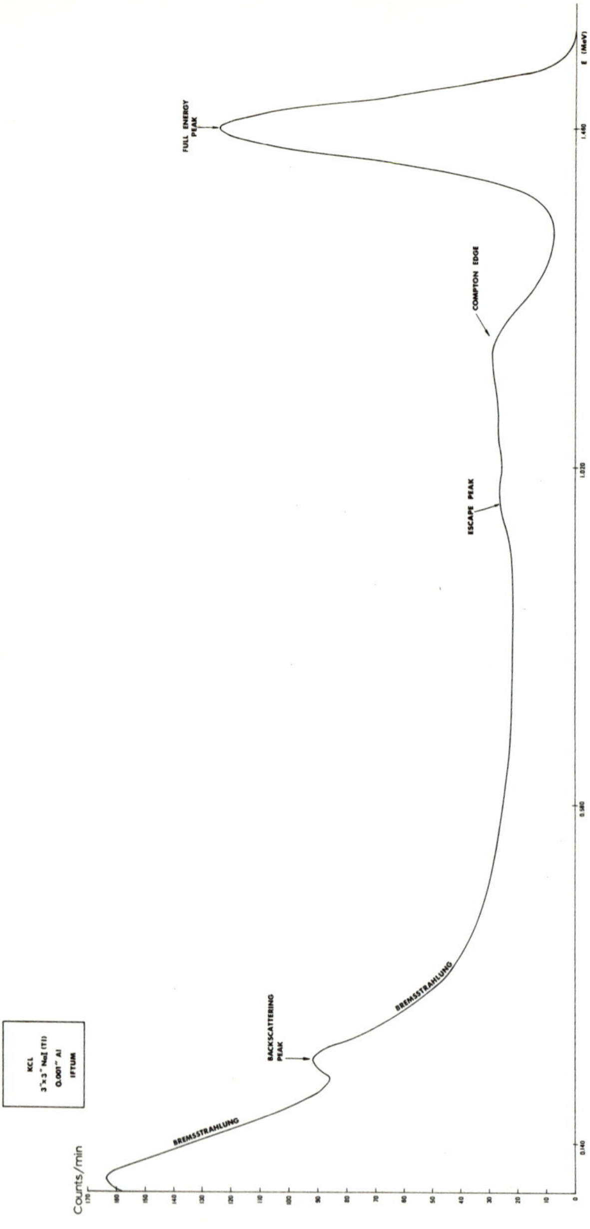

Fig.103. Gamma spectrum of ^{40}K.

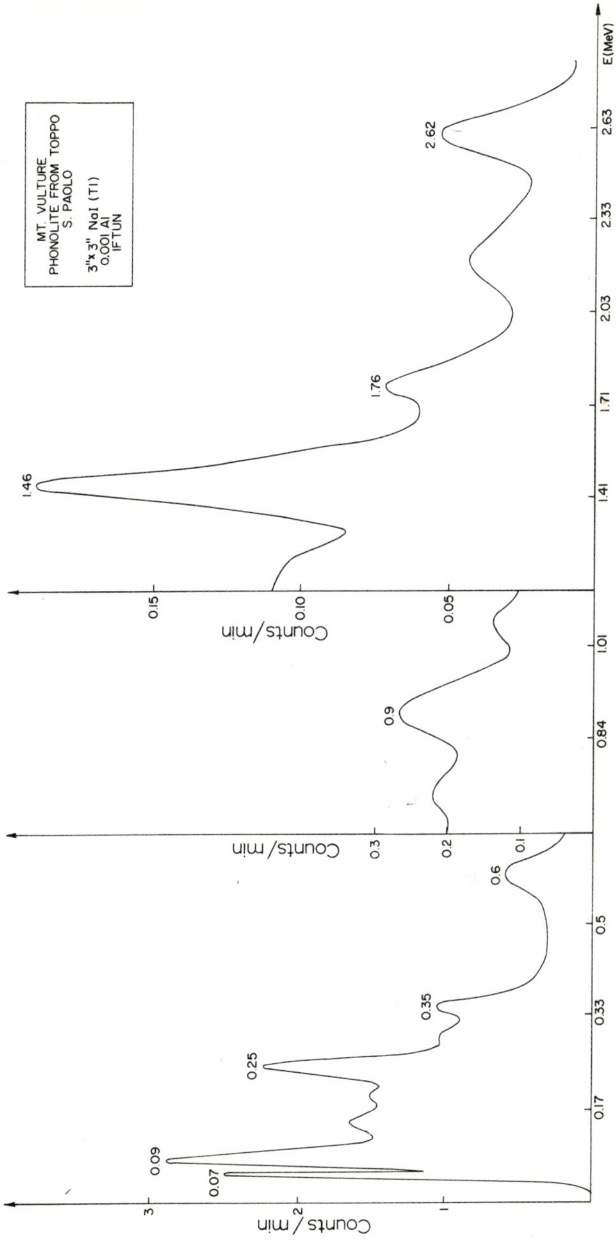

Fig.104. An ordinary rock gamma spectrum. (After BONASIA and GASPARINI, 1965.)

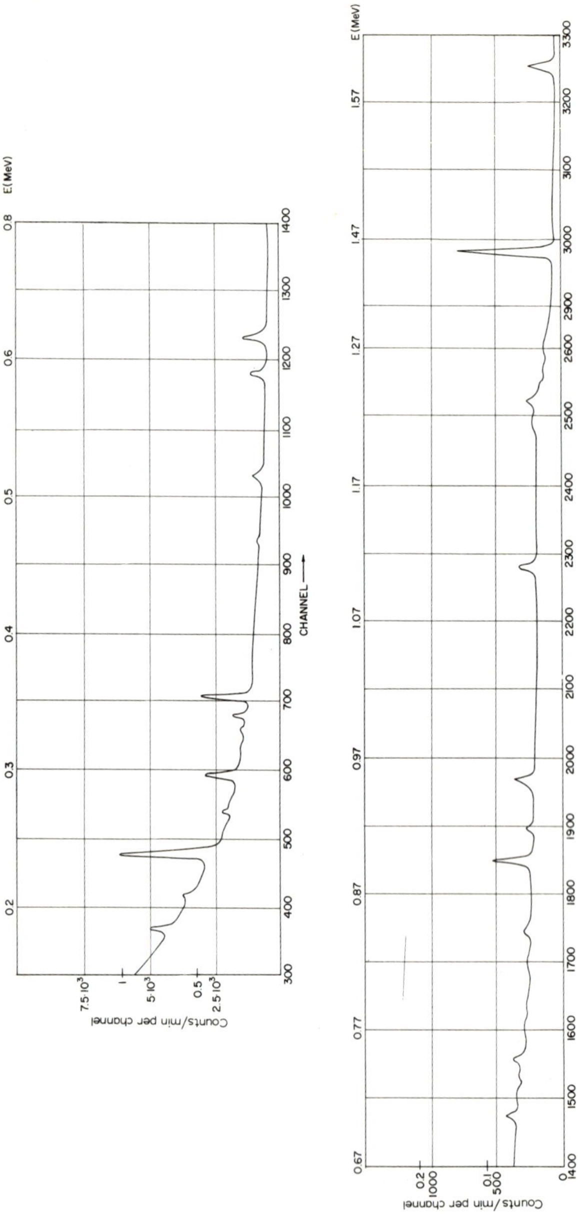

Fig.105. A Ge (Li) spectrum of a common rock. (After MATHEVON et al., 1967.)

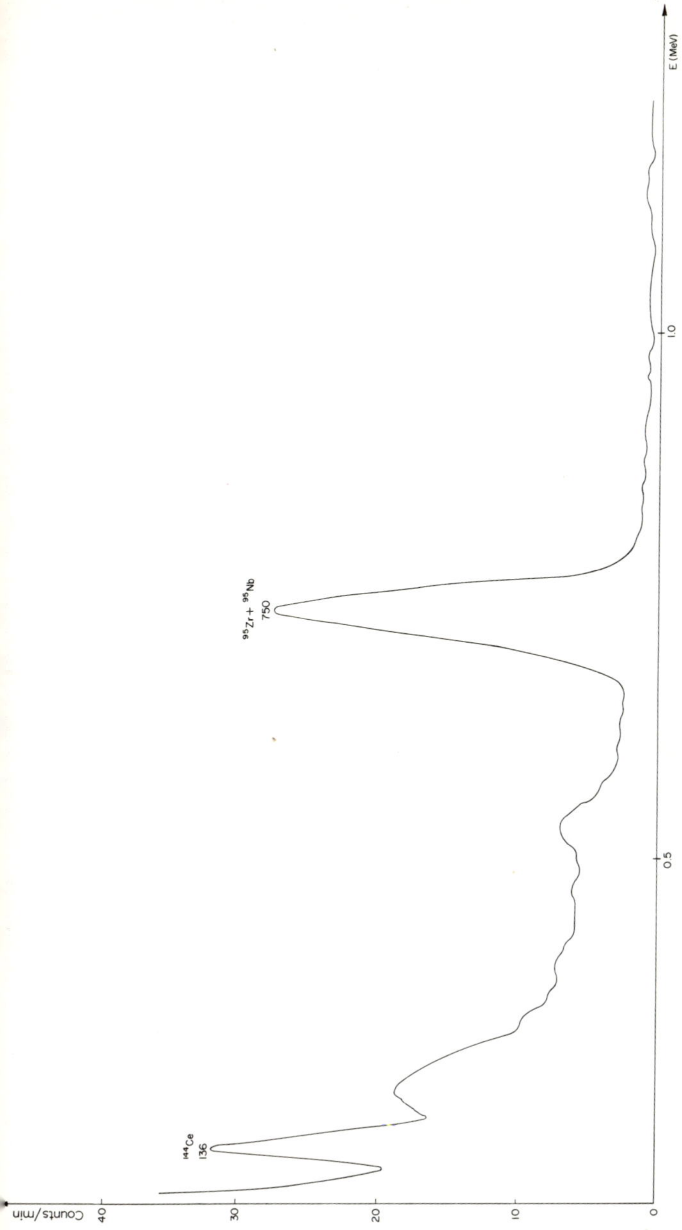

Fig.106. Fall-out spectrum obtained from a soil sample collected at the bottom of the Solfatara crater (Naples) in the spring of 1963. (After GASPARINI et al., 1963.)

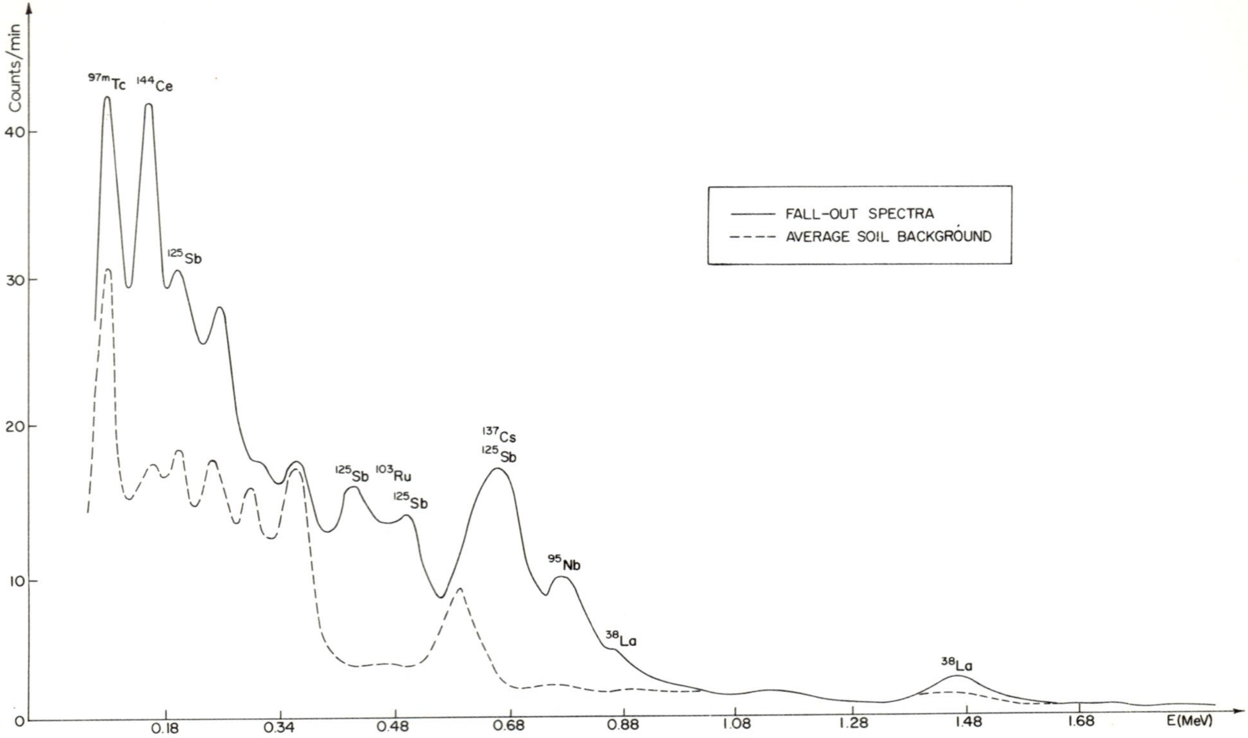

Fig.107. Fall-out spectrum from a soil sample collected at the bottom of the Solfatara crater (Naples) in the spring of 1965. (After GASPARINI and LUONGO, 1965.)

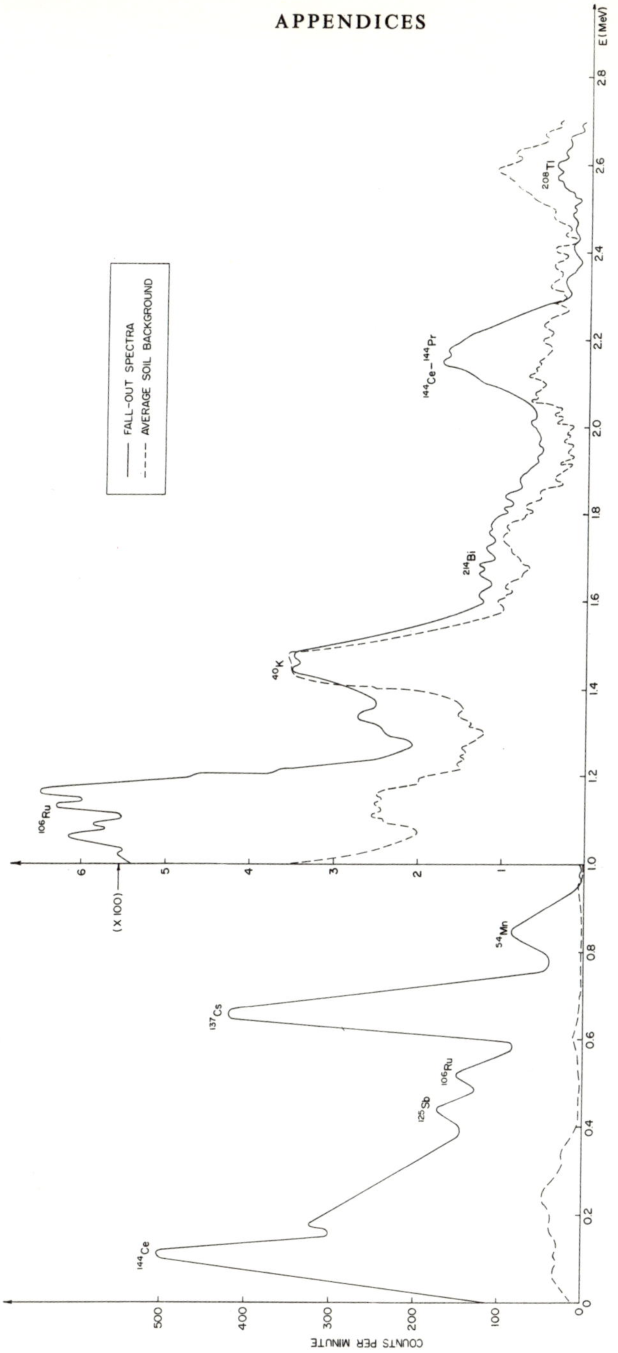

Fig.108. Fall-out spectrum from a soil sample collected at the campus of Rice University in the spring of 1965. (After ADAMS, 1966.)

APPENDIX 3

Determinations of U, Th, K_2O in Interlaboratory Geochemical Rock Standards

Data of G-1 and W-1 U.S.G.S. Standards from STEVENS (1950); SCHLECHT and STEVENS (1951); DENNEN et al. (1951); FLEISCHER and STEVENS (1962); and FLEISCHER (1965,1969).

Data of G-2, BCR-1, AGV-1, DTS-1, PCC-1, GSP-1 U.S.G.S. Standards from FLANAGAN (1969).

Data of CAAS standards from CAAS report (1961), WEBBER (1965), and SINE et al. (1969).

Data of C.R.P.G., Nancy, samples from ROUBAULT et al. (1956) and GALE (1967).

Data of G.S. of Tanganyika standard from geological survey of Tanganyika reports of 1961, 1963.

Reporters of individual determinations are quoted in most cases: for exact reference, the reader is referred to the above papers.

The following symbols have been used for the methods of analysis: AA: atomic absorption spectrophotometry; Color: colorimetric; ID: isotope dilution; MS: spark source mass spectrometry; Fluor: fluorimetry; γ spec: gamma-ray spectrometry; NA: neutron activation; Del NA: delayed neutron activation; Fl Phot: flame photometry; Chem: chemical methods; Spet: spectrophotometry; XRF: X-ray fluorescence; WR Chem: K_2O determinations in complete rock analysis.

All numbers in superscript parenthesis indicate the number of measurements averaged.

G-1 Granite from Westerly, Rhode Island (U.S.A.)

U (p.p.m.)	*Method*	*Reporter*	*Th (p.p.m.)*	*Method*	*Reporter*	K_2O *(%)*	*Method*	*Reporter*
3.8$^{(46)}$	Fluor.	Ahrens, Fleischer (1960)	45$^{(15)}$	Chem.	Ahrens, Fleischer (1960)	5.48$^{(34)}$	WR chem.	Dennen et al. (1951)
3.55$^{(2)}$	NA	Ahrens, Fleischer (1960)	51$^{(14)}$	Chem.	Ahrens, Fleischer (1960)	5.35$^{(25)}$	WR chem.	Stevens, Niles (1960)
3.2$^{(17)}$	Fluor.	Ahrens, Fleischer (1960)	52$^{(6)}$	Chem.	Ahrens, Fleischer (1960)	5.44$^{(19)}$	Fl. phot.	Stevens, Niles (1960)
3.9$^{(4)}$	γ spec.	Ahrens, Fleischer (1960)	61	XRF	Ahrens, Fleischer (1960)	5.51	Fl. phot.	Elmore (1959)
4.2$^{(29)}$	Fluor.	Grimaldi (1962)	51	Color	Ahrens, Fleischer (1960)	5.54	Fl. phot.	Brennock (1959)
3.46$^{(9)}$	NA	Das, Meinke (1960)	52$^{(57)}$	Chem.	Grimaldi (1962)	5.52$^{(19)}$	Fl. phot.	Ando et al. (1959)
2.7$^{(2)}$	NA	Hamaguchi et al. (1961)	54	Color	Abbey (1964)	5.58$^{(9)}$	Fl. phot.	Hugi (1960)
4	XRF	Ball (1964)	45.4$^{(20)}$	Spet.	May, Jenkins (1965)	5.18	Fl. phot.	Vainshtein, Lebedev (1961)
2.5 (±1.3)	α count	Cherry (1963)	61	XRF	Ball (1964)	5.47$^{(6)}$	Fl. phot.	Willgallis (1961)
3.9 (±0.8)	γ spec.	Heier (1962)	63	XRF	Filby (1964)	5.3$^{(6)}$	Spet.	Schwander (1960)
3.1	NA	Morgan, Lovering (1963)	58 (±5)	α count	Cherry (1963)	7.4	Spet.	Shaw et al. (1960)
4.70$^{(4)}$*	NA	Bate, Huizenga (1963)	51 (±2)	γ spec.	Heier (1962)	5.53$^{(8)}$	NA	Winchester (1961)
1.3	Mass spec.	Brown, Wolenstholie (1964)	42	NA	Morgan, Lovering (1963)	5.67	γ spec.	Adams (1961)
3.5	Mass spec.	Taylor (1965)	6.5	Mass spec.	Brown et al. (1964)	5.48$^{(13)}$	WR chem.	Various (1962–1965)
4	XRF	Ball (1966)	40.0	Mass spec.	Taylor (1965)	5.52	Fl. phot.	Ingamells (1962)[2]
3.7 (±0.5)	α count	Ahrens et al. (1967)	21	Spec. phot.	Tennant, Fellows (1967)	5.54	Fl. phot.	Matsui (1963)
2.16$^{(2)}$	NA	Lovering (1965)	61	XRF	Ball (1966)	5.56$^{(6)}$	Fl. phot.	Cooper (1963)
2.8$^{(3)}$	NA	Alian, Parthasarathy (1966)	52	XRF	Gunn (1967)	5.32$^{(12)}$	Fl. phot.	Biagi et al. (1963)
4.42$^{(6)}$	Del. NA	Gale (1967)	54$^{(3)}$	NA	Osawa et al. (1966)	5.46	Fl. phot.	Dalrymple, Lamphere (1964)
			80 (±2)	NA	Yule (1966)	5.48	XRF	Ingamells, Suhr (1963)[2]
			46.9$^{(6)}$	NA	Alian, Shabana (1967)	5.60	ID	Cooper (1963)
			46 (±1.5)	NA	Cobb (1967)	5.52$^{(14)}$	WR chem.	Various (1965–1969)
			53.0$^{(6)}$	Del. NA	Gale (1967)	5.53$^{(13)}$	Fl. phot.	Panto (1965)
			53.5 (±2)	α count	Ahrens (1967)	5.47$^{(20)}$	Fl. phot.	Govindaraju (1960)
			36$^{(2)}$	Mass spec.	Taylor (1965)	5.52$^{(2)}$	Fl. phot.	Ingamells et al. (1967)[2]
						5.48	XRF	Ingamells et al. (1967)[2]
						5.38	AA	Althaus (1966)
						5.52$^{(7)}$	AA	Belt (1967)
						5.06	NA	Gordon et al. (1967, 1968)
Recommended value: 4 p.p.m.			Recommended value: 52 p.p.m.			Recommended value: 5.48%		

* Contamination by ^{106}Ru.

[1] For notes on abbreviations and references, see p.244.

[2] See FLEISCHER, 1969.

TABLE XXXIV

U.S.G.S. GEOCHEMICAL STANDARDS[1]

W-1 Triassic Diabase from Fairfax County, Virginia (U.S.A.)

U (p.p.m.)	*Method*	*Reporter*	*Th (p.p.m.)*	*Method*	*Reporter*	*K_2O (%)*	*Method*	*Reporter*
$0.52^{(40)}$	Fluor.	Ahrens, Fleischer (1960)	2.6	Chem.	Ahrens, Fleischer (1960)	$0.65^{(30)}$	WR chem.	Dennere et al. (1951)
$0.53^{(2)}$	NA	Ahrens, Fleischer (1960)	$1.9^{(9)}$	Chem.	Ahrens, Fleischer (1960)	$0.64^{(28)}$	WR chem.	Various (1960)
$0.51^{(13)}$	Fluor.	Ahrens, Fleischer (1960)	$2.2^{(5)}$	Chem.	Ahrens, Fleischer (1960)	$0.65^{(17)}$	Fl. phot.	Various (1960)
$0.50^{(8)}$	γ spec.	Ahrens, Fleischer (1960)	$2.2^{(5)}$	Chem.	Ahrens, Fleischer (1960)	$0.63^{(7)}$	WR chem.	Various (1961)
$0.6^{(4)}$	Fluor.	Grimaldi (1961)	$2.1^{(4)}$	Color	Ahrens, Fleischer (1960)	$0.66^{(6)}$	Fl. phot.	Ando et al. (1959)
$0.54^{(10)}$	NA	Das, Meinke (1960)	3.6	XRF	Ahrens, Fleischer (1960)	$0.65^{(2)}$	Fl. phot.	Hugi (1960)
0.28	NA	Hamaguchi et al. (1960)	$2.3^{(14)}$	Chem.	Grimaldi (1961)	0.65	Fl. phot.	Vainshtein, Lebeden (1961)
<1	XRF	Ball (1964)	4	Color	Abbey (1964)	$0.63^{(6)}$	Fl. phot.	Willgallis (1961)
0.9 (±0.6)	α count	Cherry (1963)	$2.5^{(39)}$	Spec. phot.	May, Jenkins (1965)	$0.67^{(7)}$	Chem.	Chodas, Eugel (1961)
0.53	γ spec.	Heier, Rogers (1963)	3	XRF	Ball (1964)	0.64	Spet.	Ahrens et al. (1960)
0.49	γ spec.	Adams (1965)	3.2	XRF	Filby (1964)	$0.631^{(10)}$	NA	Winchester (1961)
0.55	NA	Morgan, Lovering (1963)	2.6 (±0.5)	α count	Cherry (1963)	0.66	γ spec.	Adams (1961)
$0.65^{(3)}$*	NA	Bate, Huizenga (1963)	4.6 (±1.3)	γ spec.	Heier (1962)	$0.63^{(13)}$	WR chem.	Various (1962–1965)
0.5	Mass spec.	Brown et al. (1964)	2.20	γ spec.	Heier, Rogers (1963)	0.62	Fl. phot.	Ingamells (1962)[2]
0.60	Mass spec.	Taylor (1965)	2.3	γ spec.	Adams (1965)	0.71	Fl. phot.	Matsui (1963)
1	XRF	Ball (1966)	2.2	NA	Morgan, Lovering (1963)	$0.66^{(4)}$	Fl. phot.	Cooper (1963)
0.48 (±0.08)	α count	Ahrens et al. (1967)	1.3	Mass spec.	Brown et al. (1964)	$0.61^{(12)}$	Fl. phot.	Biagi et al. (1963)
$0.62^{(5)}$	Del. NA	Gale (1967)	1.6	Mass spec.	Taylor (1965)	0.64	XRF	Ingamells, Suhr (1963)[2]
0.46	Mass spec.	Nicholls et al. (1967)	3	XRF	Ball (1966)	$0.67^{(10)}$	NA	Vincent, Peel (1963)
			5	XRF	Carmichael (1967)	0.65	ID	Cooper (1963)
			$2.52^{(6)}$	NA	Alian, Shabana (1967)	0.63	γ spec.	Heier, Rogers (1963)
			1.8 (±0.4)	NA	Cobb (1967)	$0.64^{(17)}$	WR chem.	Various (1965–1969)
			2.6	NA	Gordon et al. (1967)	$0.63^{(9)}$	Fl. phot.	Panto (1965)
			$3.5^{(6)}$	Del. NA	Gale (1967)	$0.65^{(20)}$	Fl. phot.	Govindaraju (1966)
			2.6 (±0.2)	α count	Ahrens et al. (1967)	0.64	Fl. phot.	Carmichael (1967)
			$2.2^{(2)}$	Mass spec.	Taylor (1965)	$0.63^{(2)}$	Fl. phot.	Ingamells et al. (1967)[2]
			2.2	Mass spec.	Nicholls et al. (1967)	0.64	XRF	Ingamells et al. (1967)[2]
						0.67	AA	Althaus (1966)
						$0.60^{(10)}$	AA	Belt (1967)
Recommended value: 0.5 p.p.m.			Recommended value: 2.4 p.p.m.			Recommended value: 0.64%		

* Contamination by ^{106}Ru

TABLE XXXV

U.S.G.S. GEOCHEMICAL STANDARDS[1]

G-2 Granite from Westerly, Rhode Island (U.S.A.)

U (p.p.m.)	*Method*	*Reporter*	*Th (p.p.m.)*	*Method*	*Reporter*	*K_2O (%)*	*Method*	*Reporter*
$1.32^{(2)}$	γ spec.	Adams (1965)	$24.5^{(2)}$	γ spec.	Adams (1965)	$3.93^{(2)}$	Fl. phot.	Majmundar (1968)
$2.16^{(2)}$	NA	Morgan, Heier (1966)	$26.6^{(2)}$	γ spec.	Bunker (1965)	$3.92^{(2)}$	γ spec.	Adams (1965)
2.1	γ spec.	Morgan, Heier (1966)	$26.1^{(2)}$	γ spec.	Heier (1965); Morgan Heier (1966)	4.50	Fl. phot.	Morgan, Heier (1966)
$1.96^{(2)}$	Chem.	Moore (1966)				4.47	γ spec.	Morgan, Heier (1966)
$2.07^{(5)}$	Del. NA	Gale (1967)	40 (est)	Del. NA	Hamilton (1966)	$4.5^{(9)}$	XRF	Van der Weijden (1966)
$1.64^{(2)}$	Del. NA	Hamilton (1966)	$26.2^{(2)}$	γ spec.	Wollenberg (1966)	$4.42^{(12)}$	Fl. phot.	Van der Weijden (1966)
1.96	γ spec.	Wollenberg (1966)	21.5	Chem.	Campbell (1967)	$4.59^{(5)}$	Fl. phot.	Van Peteghem, Ott (1966)
1.91	γ spec.	Wollenberg (1966)	$24.6^{(5)}$	Del. NA	Gale (1967)	$4.32^{(2)}$	γ spec.	Wollenberg (1966)
1.94	ID	Doe et al. (1967)	24.3	ID	Doe et al. (1967)	$4.64^{(2)}$	XRF	Joski (1967)
2.07	NA	Gordon et al. (1967, 1968)	25.9	NA	Gordon et al. (1967, 1968)	4.70	Chem.	Natl. Insti. Metal (1967)
3.4	Fluor.	Pleizier (1968)	24.9	NA	Gordon et al. (1967, 1968)	4.40	XRF	Dumesnil, Perrault (1967)
			30	XRF	Carmichael (1967)	4.8	NA	Gordon et al. (1967, 1968)
			25	XRF	Price (1967)	4.40	Fl. phot.	Siegers, Wolff (1967)
			22	Photom.	Abbey (1968)	$4.64^{(12)}$	NA	Sijperda (1968)
			24	ID	Tatsumoto (1968)	$4.46^{(2)}$	XRF	Richter (1968)
			26	NA	Greenland (1968)	$4.46^{(2)}$	XRF	Blattner (1968)
						4.48	Fl. phot.	Blattner (1968)
						4.34	Fl. phot.	Sen Gupta (1968)[2]
						4.49	Fl. phot.	Abbey (1968)
						$4.54^{(2)}$	AA	Abbey (1968)
						$4.53^{(6)}$	NA	Das (1968)
						$4.30^{(2)}$	Fl. phot.	Majmundar (1968)
						$4.55^{(30)}$	WR chem.	Various
Average value: 1.99 p.p.m. (FLANAGAN, 1969)			Average value: 25.2 p.p.m. (FLANAGAN, 1969)			Average value: 4.51%		

[1] For notes on abbreviations and references see p.244.

[2] See FLANAGAN, 1969.

TABLE XXXVI

U.S.G.S. GEOCHEMICAL STANDARDS[1]

GSP-1 Granodiorite from Silver Plume, Colorado (U.S.A.)

U (p.p.m.)	*Method*	*Reporter*	*Th (p.p.m.)*	*Method*	*Reporter*	*K_2O (%)*	*Method*	*Reporter*
0.19	γ spec.	Adams (1965)	$99.6^{(2)}$	γ spec.	Adams (1965)	$2.62^{(2)}$	Fl. phot.	Majmundar (1968)
1.7	γ spec.	Morgan, Heier (1966)	$99^{(2)}$	γ spec.	Bunker (1965)	$4.34^{(2)}$	γ spec.	Adams (1965)
$2.2^{(2)}$	Chem.	Moore (1966)	106.6	γ spec.	Heier (1965); Heier Morgan (1966)	5.50	Fl. phot.	Morgan, Heier (1966)
$2.68^{(2)}$	Del. NA	Gale (1967)				5.54	γ spec.	Morgan, Heier (1966)
$1.80^{(2)}$	Del. NA	Hamilton (1966)	100 (est)	Del. NA	Hamilton (1966)	$5.5^{(9)}$	XRF	Van der Weijden (1966)
1.11	γ spec.	Wollenberg (1966)	$113^{(2)}$	γ spec.	Wollenberg (1966)	$5.51^{(9)}$	Fl. phot.	Van der Weijden (1966)
0.27	γ spec.	Wollenberg (1966)	100	Chem.	Campbell (1967)	$5.55^{(5)}$	Fl. phot.	Van Peteghem, Ott (1966)
2.4	ID	Peterman et al. (1967)	$116.4^{(5)}$	Del. NA	Gale (1967)	$5.60^{(2)}$	γ spec.	Wollenberg (1960)
2.68	NA	Gordon et al. (1967, 1968)	106	ID	Peterman et al. (1967)	$5.60^{(2)}$	XRF	Joski (1967)
			125	NA	Gordon et al. (1967, 1968)	5.68	Chem.	Natl. Inst. Metal (1967)
			114	NA	Gordon et al. (1967, 1968)	5.25	XRF	Dumesnil, Perrault (1967)
			110	XRF	Carmichael (1967)	5.4	NA	Gordon et al. (1967, 1968)
			125	XRF	Price (1967)	5.51	Fl. phot.	Siegers, Wolff (1967)
			101	XRF	Gunn (1967a)	$5.46^{(2)}$	XRF	Blattner (1968)
			120	NA	Greenland (1968)	5.58	Fl. phot.	Blattner (1968)
						5.40	Fl. phot.	Bouvier (1968)
						5.36	Fl. phot.	Sen Gupta (1968)[2]
						5.53	Fl. phot.	Abbey (1968)
						$5.52^{(2)}$	AA	Abbey (1968)
						$5.20^{(2)}$	Fl. phot.	Majmundar (1968)
						$5.55^{(30)}$	WR chem.	Various
Average value: 1.98 p.p.m. (FLANAGAN, 1969)			Average value: 110.6 p.p.m. (FLANAGAN, 1969)			Average value: 5.48_8%		

[1] For notes on abbreviations and references see p.244.
[2] See FLANAGAN, 1969.

AGV-1 Andesite from Guano Valley, Lake County, Oregon (U.S.A.)

U (p.p.m.)	*Method*	*Reporter*	*Th (p.p.m.)*	*Method*	*Reporter*	*K_2O (%)*	*Method*	*Reporter*
1.40$^{(2)}$	γ spec.	Adams (1965)	6.44$^{(2)}$	γ spec.	Adams (1965)	4.28$^{(2)}$	Fl. phot.	Majmundar (1968)
2.18$^{(2)}$	NA	Morgan, Heier (1966)	6.2$^{(2)}$	γ spec.	Bunker (1965)	2.62$^{(2)}$	γ spec.	Adams (1965)
1.81$^{(2)}$	γ spec.	Bunker (1965)	6.52$^{(2)}$	γ spec.	Heier (1965); Morgan Heier (1966)	2.91	XRF	Lewis (1966)
1.9	γ spec.	Morgan, Heier (1966)				2.94	Fl. phot.	Morgan, Heier (1966)
1.96$^{(2)}$	Chem.	Moore (1966)	10 (est)	Del. NA	Hamilton (1966)	2.93	γ spec.	Morgan, Heier (1966)
2.09$^{(5)}$	Del. NA	Gale (1967)	6.66$^{(2)}$	γ spec.	Wollenberg (1966)	3.0$^{(9)}$	XRF	Van der Weijden (1966)
1.46$^{(2)}$	Del. NA	Hamilton (1966)	6.06	Chem.	Campbell (1967)	2.92$^{(9)}$	Fl. phot.	Van der Weijden (1966)
2.18	γ spec.	Wollenberg (1966)	8.1$^{(5)}$	Del. NA	Gale (1967)	2.93$^{(5)}$	Fl. phot.	Van Peteghem, Ott. (1966)
1.83	γ spec.	Wollenberg (1966)	7.0	NA	Gordon et al. (1967, 1968)	2.82$^{(2)}$	γ spec.	Wollenberg (1966)
2.09	NA	Gordon et al. (1967, 1968)	5.9	NA	Gordon et al. (1967, 1968)	2.81$^{(8)}$	XRF	Sine (1967)
2.6	Fluor.	Pleizier (1968)	10	XRF	Carmichael (1967)	3.19$^{(2)}$	XRF	Joski (1967)
1.96$^{(5)}$	ID	Tatsumoto (1968)	5	XRF	Price (1967)	3.03	Chem.	Natl. Inst. Metal (1967)
			7.0	XRF	Gunn (1967)	2.80	XRF	Dumesnil, Perrault (1967)
			8	Photom.	Abbey (1968)	2.6	NA	Gordon et al. (1967, 1968)
			6.27$^{(5)}$	ID	Tatsumoto (1968)	2.96	Fl. phot.	Siegers, Wolff (1967)
			6.6	NA	Greenland (1968)	2.95$^{(2)}$	XRF	Richter (1968)
						2.86$^{(2)}$	XRF	Blattner (1968)
						2.95	Fl. phot.	Blattner (1968)
						2.89	Fl. phot.	Sen Gupta (1968)[2]
						2.94	Fl. phot.	Bouvier (1968)
						2.80	Fl. phot.	Sen Gupta (1968)[2]
						2.90	Fl. phot.	Abbey (1968)
						2.92$^{(2)}$	AA	Abbey (1968)
						2.9	XRF	Brennock (1968)
						2.40$^{(2)}$	Fl. phot.	Majmundar (1968)
						2.93$^{(29)}$	WR chem.	Various
Average value: 1.94 p.p.m. (FLANAGAN, 1969)			Average value: 6.96 p.p.m. (FLANAGAN, 1969)			Average value: 2.89_8%		

[1] For notes on abbreviations and references see p.244.
[2] See FLANAGAN, 1969.

TABLE XXXVIII

U.S.G.S. GEOCHEMICAL STANDARDS[1]

BCR-1 Basalt from Columbia River (Yakima Type), Washington, Oregon (U.S.A.)

U (p.p.m.)	*Method*	*Reporter*	*Th (p.p.m.)*	*Method*	*Reporter*	*K_2O (%)*	*Method*	*Reporter*
$1.20^{(2)}$	γ spec.	Adams (1965)	$6.09^{(2)}$	γ spec.	Adams (1965)	$3.23^{(8)}$	Fl. phot.	Majmunder (1968)
$1.81^{(2)}$	NA	Morgan, Heier (1965)	$6.1^{(2)}$	γ spec.	Bunker (1965)	1.46	γ spec.	Adams (1965)
$2.09^{(2)}$	γ spec.	Bunker (1965)	$6.13^{(2)}$	γ spec.	Heier (1965); Morgan Heier (1966)	1.79	XRF	Lewis (1966)
1.6	γ spec.	Morgan, Heier (1966)				1.70	Fl. phot.	Morgan, Heier (1966)
$1.6^{(2)}$	Chem.	Moore (1966)	10 (est)	Del. NA	Hamilton (1966)	1.67	γ spec.	Morgan, Heier (1966)
$1.80^{(4)}$	Del. NA	Gale (1967)	6.62	γ spec.	Wollenberg (1966)	$1.75^{(9)}$	XRF	Van der Weijden (1966)
$1.42^{(2)}$	Del. NA	Hamilton (1966)	6.20	Chem.	Campbell (1967)	$1.70^{(9)}$	Fl. phot.	Van der Weijden (1966)
2.17	γ spec.	Wollenberg (1966)	$7.6^{(4)}$	Del. NA	Gale (1967)	$1.75^{(5)}$	Fl. phot.	Van Peteghem, Ott. (1966)
1.68	γ spec.	Wollenberg (1966)	6.7	NA	Gordon et al. (1967, 1968)	$1.59^{(2)}$	γ spec.	Wollenberg (1966)
1.80	NA	Gordon et al. (1967, 1968)	15	XRF	Carmichael (1967)	$1.63^{(8)}$	XRF	Sine (1967)
2.2	Fluor.	McNail (1968)	<5	XRF	Price (1967)	$2.02^{(2)}$	XRF	Joski (1967)
$1.73^{(5)}$	ID	Tatsumoto (1968)	5.5	XRF	Gunn (1967)	1.79	Chem.	Natl. Inst. Metal. (1967)
			7	Photom.	Abbey (1968)	1.56	XRF	Dumesnil, Perrault (1967)
			$5.99^{(5)}$	ID	Tatsumoto (1968)	1.4	NA	Gordon et al. (1967, 1968)
			5.8	NA	Greenland (1968)	1.70	Fl. phot.	Siegers, Wolff (1967)
						$1.72^{(2)}$	XRF	Richter (1968)
						$1.70^{(2)}$	XRF	Blattner (1968)
						$1.69^{(2)}$	Fl. phot.	Blattner (1968)
						1.69	Fl. phot.	Sen Gupta (1968)[2]
						1.62	Fl. phot.	Sen Gupta (1968)[2]
						1.65	Fl. phot.	Abbey (1968)
						$1.69^{(2)}$	AA	Abbey (1968)
						1.6	XRF	Brennock (1968)
						$1.64^{(8)}$	Fl. phot.	Majmundar (1968)
						$1.65^{(29)}$	WR chem.	Various
Average value: 1.73 p.p.m.			Average value: 6.81 p.p.m.			Average value: 1.68_4 %		

[1] For notes on references and abbreviations see p.244.

[2] [illegible]

TABLE XXXIX

U.S.G.S. GEOCHEMICAL STANDARDS[1]

PCC-1 Periodotite collected as stream boulders from Caradero ultramafic mass, Sonoma County, California (U.S.A)

U (p.p.m.)	*Method*	*Reporter*	*Th (p.p.m.)*	*Method*	*Reporter*	*K_2O (%)*	*Method*	*Reporter*
0.0040[(2)]	NA	Morgan, Heier (1966)	<5	XRF	Carmichael (1967)	0.00	Fl. phot.	Majmundar (1968)
0.005	Del. NA	Gale (1967)	0.0108[(2)]	NA	Hamaguchi; Omuna (1967)	0.00138	NA	Morgan, Heier (1966)
~0.005	Del. NA	Hamilton (1966)	5	XRF	Price (1967)	<0.1	XRF	Van der Weijden (1966)
0.005	NA	Gordon et al. (1967, 1968)				0.0012[(3)]	Fl. phot.	Van Peteghem, Ott. (1966)
0.0049	NA	Hamaguchi; Omuna (1962)				0.09[(8)]	XRF	Sine (1967)
						0.36[(2)]	XRF	Joski (1967)
						0.00	XRF	Dumesnil, Perrault (1967)
						0.006	Fl. phot.	Siegers, Wolff (1967)
						<0.005	XRF	Richter (1968)
						0.05[(9)]	WR chem.	Various
						0.0	XRF	Blattner (1968)
						0.0	Fl. phot.	Blattner (1968)
						0.0013	NA	Wakita et al. (1967)
						0.01	Fl. phot.	Sen Gupta (1968)[2]
						<0.01	AA	Abbey (1968)
						0.00	Fl. phot.	Majmundar (1968)
Average value: 0.0048 p.p.m.			Average value: 0.018 p.p.m.			Average value: 0.019%		

[1] For notes on abbreviations and references see p.244.

[2] See FLANAGAN, 1969.

TABLE XL

U.S.G.S. GEOCHEMICAL STANDARDS[1]

DTS-1 Dunite from Twin Sisters Area, Hamilton, Washington (U.S.A.)

U (p.p.m.)	*Method*	*Reporter*	*Th (p.p.m.)*	*Method*	*Reporter*	*K_2O (%)*	*Method*	*Reporter*
0.0032$^{(2)}$	NA	Morgan, Heier (1966)	<5	XRF	Carmichael (1967)	0.00	Fl. phot.	Majmunder (1968)
0.004$^{(4)}$	Del. NA	Gale (1967)	0.0092$^{(2)}$	NA	Hamaguchi; Omuna (1967)	0.00106	NA	Morgan, Heier (1966)
~0.003	Del. NA	Hamilton (1966)	<5	XRF	Price (1967)	<0.1	XRF	Van der Weijden (1966)
0.004	NA	Gordon et al. (1967; 1968)	<1	XRF	Gunn (1967)	0.0026$^{(3)}$	Fl. phot.	Van Peteghem, Ott. (1966)
0.0040	NA	Hamaguchi; Omuna (1967)				0.40$^{(2)}$	XRF	Joski (1967)
						0.00	XRF	Dumesnil, Perrault (1967)
						0.014	Fl. phot.	Siegers, Wolff (1967)
						<0.05	XRF	Richter (1968)
						0.0	XRF	Blattner (1968)
						0.0	Fl. phot.	Blattner (1968)
						0.0011	NA	Wakita et al. (1967)
						0.01	Fl. phot.	Sen Gupta (1968)[2]
						<0.01	Fl. phot.	Abbey (1968)
						<0.01	AA	Abbey (1968)
						0.05$^{(12)}$	WR chem.	Various
Average value: 0.0037 p.p.m.			Average value: 0.0092 p.p.m.			Average value: 0.02_3 %		

[1] For notes on abbreviations and references see p.244.

[2] See FLANAGAN, 1969.

U (p.p.m.)	Method	Reporter	Th (p.p.m.)	Method	Reporter	K_2O (%)	Method	Reporter
2350[4]	–	CAAS (1961)	1270[3]	–	CAAS (1961)	0.62[3]	–	CAAS (1961)
1900	Fluor.	Webber (1965)	610	Spet.	Webber (1965)	2.39	Fl. phot.	Webber (1965)
2230	Fluor.	Webber (1965)	1000	Spet.	Webber (1965)	2.5	Spet.	Webber (1965)
2395 (±47)	α count	Webber (1965)	1230	–	Webber (1965)	2.59	Chem.	Webber (1965)
2395	–	Webber (1965)	1400	–	Webber (1965)	2.60	Chem.	Webber (1965)
2400	–	Webber (1965)	1500	Spet.	Webber (1965)	2.63	Fl. phot.	Webber (1965)
2400	–	Webber (1965)	1731(±55)	γ spec.	Webber (1965)	2.65	Fl. phot.	Webber (1965)
~2500	Spet.	Webber (1965)	1800	Spet.	Webber (1965)	2.65	–	Webber (1965)
2500	Fluor.	Webber (1965)	1100	XRF	Ball (1968)	2.69	Fl. phot.	Webber (1965)
2600	–	Webber (1965)	1375	Spet.	Moxham (1967)	2.71	XRF	Webber (1965)
2868 (±114)	γ spec.	Webber (1965)	1300	Spet.	Tennant, Fellows (1964)	2.76	Fl. phot.	Webber (1965)
3000	Spet.	Webber (1965)				2.8	XRF	Webber (1965)
2690(±155)	Fluor.	Pajer et al. (1965)				2.80	–	Webber (1965)
2700	Spet.	Tennant, Fellow (1967)				2.89	Fl. phot.	Webber (1965)
2700	XRF	Ball (1968)				2.94	–	Webber (1965)
						2.98	Fl. phot.	Webber (1965)
						3.34	γ spec.	Webber (1965)
						2.35	–	Gustafson (1965)
						2.41	XRF	Hahn, Weinheimer (1955)
						2.44	XRF	Ball (1968)
						2.54	Fl. phot.	Van Peteghem (1968)
						2.56	Fl. phot.	Bloxam et al. (1964)
						2.56	Fl. phot.	Henderson (1967)
						2.60	XRF	Kodama et al. (1964)
						2.61	Fl. phot.	Bloxam et al. (1964)
						2.62	Fl. phot.	Pajari et al. (1965)
						2.62	Fl. phot.	Vijan et al. (1968)
						2.63	XRF	Van der Weijden (1966)
						2.66	Fl. phot.	Bloxam et al. (1964)
						2.70	XRF	Richter (1968)
						2.75	XRF	Condie (1966)
						2.84	Fl. phot.	Hosking (1965)
Mean value: 2521			Mean value: 1305			Mean value: 2.67		
Median: 2500			Median: 1338			Median: 2.63		

[1] For notes on abbreviations and references see p.244.

TABLE XLII

CANADIAN ASSOCIATION FOR APPLIED SPECTROSCOPY (CAAS) STANDARDS[1]

Sulphide ore–1—from Falconbridge, Ontario (Canada)

U (p.p.m.)	*Method*	*Reporter*	*Th (p.p.m.)*	*Method*	*Reporter*	*K_2O (%)*	*Method*	*Reporter*
<500	Spect.	Webber (1965)	<300	Spet.	Webber (1965)	0.62(3)	Chem.	CAAS (1961)
1.12 (±0.04)	γ spec.	Webber (1965)	3.43 (±0.17)	γ spec.	Webber (1965)	0.64(7)	WR chem.	Webber (1965)
1.7	–	Webber (1965)	1	Spet.	Cohen et al. (1968)	0.55	Fl. phot.	Webber (1965)
25.9	Fluor.	Webber (1965)				0.53 (±0.012)	–	Webber (1965)
0.97 (±0.17)	Fluor.	Perry (1966)				0.61	Chem.	Webber (1965)
						0.61	Fl. phot.	Webber (1965)
						0.62	Fl. phot.	Webber (1965)
						0.64	XRF	Webber (1965)
						0.70	–	Webber (1965)
						0.72	–	Webber (1965)
						0.54	XRF	Hahn, Weinhemer (1965)
						Mean value:	0.62	
						Median:	0.61	

[1] For notes on abbreviations and references see p.244.

TABLE XLIII

ROCK STANDARDS DISTRIBUTED FROM THE "CENTRE DES RECHERCHES PETROGRAPHIQUES ET GEOCHIMIQUE" NANCY (FRANCE)[1]

U (p.p.m.)	*Method*	*Reference*	*Th (p.p.m.)*	*Method*	*Reference*	*K_2O (%)*	*Method*	*Reference*
4.26 (±0.04)[4]	Del. NA	Gale (1967)	15.8 (±2.7)[4]	Granite—GA Del. NA	Gale (1967)	4.05	Chem. spet.	Roubault et al. (1964)
21.34 (±0.15)[4]	Del. NA	Gale (1967)	45.0 (±5.0)	Granite—GR Del. NA	Gale (1967)	4.50	Chem, spet.[14]	Roubault et al. (1964)
19.71 (±0.28)[4]	Del. NA	Gale (1967)	111.0 (±5.0)[4]	Granite—GH Del. NA	Gale (1967)	4.70	Chem. spet.	Roubault et al. (1964)
2.65 (±0.01)[4]	Del. NA	Gale (1967)	12.6 (±1.2)[4]	Basalt—BR Del. NA	Gale (1967)	1.15	Chem. spet.	Roubault et al. (1964)

[1] For notes on abbreviations and references see p.244.

TABLE XLIV

GEOLOGICAL SURVEY OF TANGANIYKA STANDARD[1]

T-1—Msusule Toualite

U (p.p.m.)	*Method*	*Reporter*	*Th (p.p.m.)*	*Method*	*Reporter*	*K_2O (%)*	*Method*	*Reporter*
0.52 (±0.04)	γ spec.	Adams	2.76 (±0.08)	γ spec.	Adams	1.20	WR chem.	GST-Report (1961)
						1.24[(14)]	WR chem.	Various (1961–1963)
						1.27	Fl. phot.	Adams
						1.21	Chem.	Martinec
						1.20 (±0.01)	γ spec.	Adams

[1] For notes on abbreviations and references see p.244.

References

ABROSIMOV, N. K. and COCHAROV, G. E., 1962. The effect of the thickness on the shape of the energy distribution curves of alpha particles. *Dokl. Akad. Nauk. S.S.S.R., Ser. Phys.*, 26: 237–244.

ADAMS, J. A. S., 1963. Nondestructive gamma spectrometry of serpentinite cores from Mayaguez, Puerto Rico. In: *A Study of Serpentinite, Natl. Acad. Sci.—Natl. Res. Council Publ.*, 1188: 145–148.

ADAMS, J. A. S., 1964. Laboratory gamma-ray spectrometer for geochemical studies. In: J. A. S. ADAMS and W. H. LOWDER (Editors), *The Natural Radiation Environment*. University of Chicago Press, Chicago, Ill., pp.485–498.

ADAMS, J. A. S., 1968. Total and spectrometric gamma-ray surveys from helicopters and vehicles. *Symp. Use Nuclear Techniques Prospecting Develop. Mineral Resources, Buenos Aires*. International Atomic Energy Agency, Vienna, 14 pp.

ADAMS, J. A. S. and FRYER, G. E., 1964. Portable gamma-ray spectrometer for field determination of thorium, uranium and potassium. In: J. A. S. ADAMS and W. M. LOWDER (Editors), *The Natural Radiation Environment*. University of Chicago Press, Chicago, Ill., pp.577–596.

ADAMS, J. A. S. and LOWDER, W. M., 1964. Editor's postscriptum. In: J. A. S. ADAMS and W. M. LOWDER (Editors), *The Natural Radiation Environment*. University of Chicago Press, Chicago, Ill., pp.1025–1027.

ADAMS, J. A. S. and MAECK, W. J., 1954. Fluorimetric and colorimetric micro-determinations of uranium in rocks and minerals. *Anal. Chem.*, 26; 1635–1639.

ADAMS, J. A. S., RICHARDSON, K. E. and TEMPLETON, C. C., 1958. Determination of thorium and uranium in sedimentary rocks by two independent methods. *Geochim. Cosmochim. Acta*, 13: 270–279.

ADAMS, J. A. S., OSMOND, J. K. and ROGERS, J. J. W., 1959. The geochemistry of thorium and uranium. In: L. H. AHRENS, F. PRESS, K. RANKAMA, S. K. RUNCORN (Editors), *Physics and Chemistry of the Earth, 3*. Pergamon, London, pp.298–348.

ADAMS, J. A. S., FRYER, G. E. and ROGERS, J. J. W., 1969. Gamma-ray spectrometry for airborne geochemistry. *Anal. Chem.*, 41:22A–28A.

AHRENS, L. H., 1951. Spectrochemical analyses of some of the rare elements in the granite and diabase samples. *U.S. Geol. Surv. Bull.*, 980: 53–58.

AHRENS, L. H. and FLEISCHER, M., 1960. Report on trace constituents in granite G-1 and diabase W-1. *U.S. Geol. Surv. Bull.*, 1113: 83–111.

ALDER, P., OESCHGER, H. and WASSON, J. T., 1967. Aluminum-26 in deep sea sediments. In: *Radioactive Dating and Methods of Low Level Counting*. International Atomic Energy Agency, Vienna, pp.189–195.

ALIAGA-KELLY, D. and NICOLL, D. R., 1966. Recent developments in scintillation detectors. *Nucl. Instr. Methods*, 43: 110–115.

AMIEL, S., 1962. Analytical application of delayed neutron emission in fissionable elements. *Anal. Chem.*, 34: 1683–1692.

AMIEL, S., GILAT, J. and HEYMANN, D., 1967. Uranium content of chrondrites by thermal neutron activation and delayed neutrons counting. *Geochim. Cosmochim. Acta*, 31: 1499–1504.

AMREHN, H. und KULENKAMPF, H., 1955. Energieverteilung im Spektrum der Röntgenbremsstrahlung dünner Antikathoden in Abhängigkeit von Ordnungzahl und Spannung. *Z. Physik*, 140: 452–464.

ANDERS, E., 1960. The record in the meteorites, II: on the presence of aluminum-26 in meteorites and tektites. *Geochim. Cosmochim. Acta*, 19: 53–62.

ANDRESEN, E. G., 1965. Die Totzeitkorrektur. *Kerntechnik*, 7: 79–81.

ANGUS, J., COCKROFT, A. L. and CURRAN, S. C., 1949. Investigation of soft radiations by proportional counters, III. The beta spectrum of carbon-14. *Phil. Mag.*, 40: 522–530.

AREVALO-CARRETTERO, P., BURRI, C. und WEIBEL, M., 1962. Zur Petrochemie des Roccamonfina-vulkans (Prov. Caserta, Italien). *Schweiz. Mineral. Petrog. Mitt.*, 42: 237–268.

ARMSTRONG, R. L., 1966. K–Ar dating using neutron activation for Ar analysis; granitic plutons of the eastern Great Basin, Nevada and Utah. *Geochim. Cosmochim. Acta*, 30: 565–600.

ARONSON, R. and FRENCH, R. L., 1962. Air, ground and structure scattering of radiation. In: L. P. BLIZARD and L. S. ABBOTT (Editors), *Reactor Handbook, 3. Part B*, 2nd ed. Interscience, New York, N.Y., pp.184–284.

ATTREE, R. W., CABELL, M. J., CUSHING, R. L. and PIERONI, J. J., 1962. A calorimetric determination of the half-life of Thorium-230 and a consequent revision of its neutron capture cross section. *Can. J. Phys.*, 40: 194–201.

AVAN, L. et KELLER, P., 1961a. Dosage de l'uranium et du thorium dans les minerais radiatifs par spectrometrie gamma. *Compt. Rend.*, 252: 1135–1137.

AVAN, L. et KELLER, P., 1961b. Dosage de l'uranium et du thorium dans les minerais radiatifs par spectrometrie gamma. *Compt. Rend.*, 252: 1445–1447.

AZMAN, A., MOLJK, A. and PAHOR, J., 1968. Electron capture in K-40. *Z. Physik*, 208: 234–237.

BACKENSTOSS, G. und GOEBEL, K., 1955. Die Zahl der Gamma-Quanten des Kaliums. *Z. Naturforsch.*, 10a: 920–930.

BALACEK, K. J., 1966. *Radioactivity in Tektites.* Thesis, Rice University, Houston, Texas.

BANKS, P. O. and SILVER, L. T., 1966. Evaluation of the decay constant of uranium-235 from lead isotope ratios. *J. Geophys. Res.*, 71: 4037–4046.

BARBEZIER, J., BLANGY, B., GUITTON, J. and LALLEMANT, J., 1958. Methods of carborne and airborne prospecting—the techniques of radiation prospecting by energy discrimination. *Proc. Conf. Peaceful Use At. Energy, 2nd, Geneva, Switzerland*, pp.799–814.

BAUMGAERTNER, F., 1967. Table of neutron activation constants. In: *Buchreihe der Atomkernenergie, 4.* Karl Thiemig, München, 115 pp.

BECK, H. L., CONDON, W. J. and LOWDER, W. M., 1964a. Environmental radiation measurements in the southeastern, central and western United States. *U.S. At. Energy Comm., H.A.S.L. Rept.*, 145.

BECK, H. L., CONDON, W. J. and LOWDER, W. M., 1964b. Spectrometric technique for measuring environmental gamma radiation. *U.S. At. Energy Comm., H.A.S.L. Rept.*, 150.

BELL, P. R., 1955. The scintillation method. In: K. SIEGBAHN (Editor), *Beta and Gamma Ray Spectroscopy*. North-Holland, Amsterdam, pp.132–164.

BERGER, M. J. and DOGGETT, J., 1956. Response function of thallium activated sodium iodide scintillation counters. *J. Res. Natl. Bur. Std.*, 56: 355–366.

BERGER, M. J. and RASO, D. J., 1960. Montecarlo calculations of gamma-ray backscattering. *Radiation Res.*, 12: 20–37.

BERTOLINI, G., BISI, A., LAZZARINI, E. and ZAPPA, L., 1954. On the L to K capture ratio in ^{109}Cd. *Nuovo Cimento*, 11: 530–550.

BERZINA, I. G. and DOLOMANOVA, E. I., 1967. Uranium content in cassiterites determined by uranium fission fragments tracks. *Dokl. Akad. Nauk S.S.S.R.*, 175: 171–174.

BIRCH, F., 1954. Heat from radioactivity. In: H. FAUL (Editor), *Nuclear Geology*. Wiley, New York, N.Y., pp.148–174.

BIRCH, F., 1965. Speculations on the thermal history of the earth. *Geol. Soc. Am. Bull.*, 76: 133–154.

BIRKS, J. B., 1953. *Scintillation Counters*. Pergamon, London, 148 pp.

BIRKS, J. B., 1964a. *The Theory and Practice of Scintillation Counting*. Pergamon, London, 662 pp.

BIRKS, J. B., 1964b. The scintillation process in alkali halide crystals. *Inst. Elec. Electron, Engrs., Trans. Nucl. Sci.*, 11: 4–11.

BLANC, D., MATHIEU, J., PATAU, J. P., FRANCOIS, H. et SAUDAIN, G., 1966. Le comportement sous irradiation gamma des chambres d'ionization cylindriques remplies d'hexane liquide. *Nucl. Instr. Methods*, 41: 272–276.

BLIZARD, E. P., 1962. Analytical methods of shield design. In: E. P. BLIZARD and L. S. ABBOTT (Editors), *Reactor Handbook, 3, Part B,* 2nd ed. Interscience, New York, N.Y., pp.128–150.

BONASIA, V. e GASPARINI, P., 1965. Contributo allo studio dell'evoluzione magmatica del Vulture mediante misure di radioattività. *Ann. Osserv. Vesuviano*, 7:44–57.

BONFIGLIOLI, G., FERRO, A. and MOJONI, A., 1961. Electron microscope investigation on the nature of fission products in mica. *J. Appl. Phys.*, 32: 2499.

BOOMAN, G. L. and REIN, J. E., 1962. Uranium. In: I. M. KOLTHOFF and P. J. ELVING (Editors), *Treatise on Analytical Chemistry, Part II, 9*. Interscience, New York, N.Y., pp.1–188.

BOWIE, S. H. U., 1954. Nuclear emulsions techniques. In: H. FAUL (Editor), *Nuclear Geology*. Wiley, New York, N.Y., pp.148–174.

BOWMAN, H. R., HYDE, E. K., THOMPSON, S. G. and JARED, R. C., 1966. Application of high resolution semiconductor detectors in X-ray emission spectrography. *Science*, 151: 562–568.

BOWYER-STUART, C., 1965. Large collecting area sealed proportional counter. *Rev. Sci. Instr.*, 36: 1009–1012.

BRADT, H., GUGELOT, P. C., HÜBER, O., MEDICUS, H., PREISWERK, P. und SCHERRER, P., 1946. Empfindlichkeit von Zählrohren mit Blei-, Messing-, und Aluminumkathode für gamma-Strahlung im Energieintervall 0.1 Mev bis 3 Mev. *Helv. Phys. Acta*, 19: 79–90.

BRINCKMANN, P., 1965. CsI(Na) scintillation crystals. *Phys. Letters*, 15: 305.

BROWN, S. C. and EVANS, R. D., 1939. Discharge counter of unconventional geometry. *Phys. Rev.*, 55: 1133.

BUNKER, C. M. and BUSH, C. A., 1967. A comparison of potassium analysis by gamma-ray spectrometry and other techniques. *U.S. Geol. Surv., Profess. Papers*, 575B: B164–B169.

BURCH, P. R. A., 1953. Specific gamma-activity, the branching ratio and half-life of potassium 40. *Nature*, 172: 361.

BYRNE, J., 1965. Calculation of the detection efficiency of a cylindrical counter for X-rays from an external source. *Nucl. Instr. Methods*, 35: 125–129.

CAMPBELL, J. L. P., 1961. Radioactivity well logging. In: G. B. MOODY (Editor), *Petroleum Exploration Handbook*. McGraw-Hill, New York, N.Y., pp.20/1–20/40.

CASERTANO, L., 1952. Sulla radioattività dei prodotti vesuviani. *Ann. Osserv. Vesuviano*, 1: 313–322.

CERRAI, I. E., SCHREIBER, B., TRIULZI, C. and TASSI PELATI, L., 1964. Contributions of ^{144}Ce, ^{144}Pr, and ^{147}Pm to the radioactivity in the surface layers of coastal sediments in the Ligurian sea. *Rend. Ist. Lombardo Sci. Lettere, B*, 98: 184–196.

CHAPMAN, G. T., 1967. Gamma-ray attenuation coefficient for germanium. *Nucl. Instr. Methods*, 52: 101–103.

CHARLET, J. M., 1968. Utilization du spectre gamma a basse energie (20–610 keV) dans la determination des mineraux radioactifs. *Bull. Soc. Franç. Mineral. Crist.*, 91: 151–165.

CHAVLOV, P. I., 1959. The ^{234}U/^{238}U ratio in some secondary minerals. *Geochemistry*, 2: 203–210.

CHEMINEE, J. L. and NORDEMANN, D., 1963. Dosage du potassium, de l'uranium, du radium et du thorium par spectrometrie gamma dans les laves de l'Auvergne, du Velay et de la province sicilienne. *Bull. Soc. Géol. France*, 5: 218–229.

CHERDYNTZEV, V. V., ORLOV, D. P., ISABAEV, E. A. and IVANOV, V. I., 1961. Uranium isotopes in nature. II. Isotopic composition of uranium minerals. *Geochemistry*, 10: 927–936

CHERDYNTZEV, V. V., KAZACHEVSKIY, I. V., SULERZITSKIY, L. D. and KUZ'MINA, YE. A., 1965. Plutonium 239 in nature. *Geochem. Intern.*, 2: 918–922.

CHERDYNTZEV, V. V., KISLIZTINA, G. I., KUPTSOV, V. M., KUZ'MINA, YE. A. and ZVEREV, V. L., 1967. Radioactivity and absolute age of young volcanic rocks. *Geochem. Intern.*, 4: 689–696.

CHERRY, R. D. and ADAMS, J. A. S., 1963. Gamma spectrometric determination of thorium, uranium and potassium in tektites. *Geochim. Cosmochim. Acta*, 27: 1089–1096.

CHOPPIN, G. R. and SIKKELAND, T., 1960. Scheme for the separation of the elements francium through uranium. In: E. K. HYDE (Editor), *Radiochemistry of Thorium. Natl. Acad. Sci.—Natl. Res. Council Nucl. Sci. Ser.*, 3004: 60–61.

CHOY, S. C. and SCHMITT, R. A., 1965. Gamma-ray spectra analyzed by computer programme using the peak area method. *Nature*, 205: 758–760.

CIVETTA, L., GASPARINI, P. e RAPOLLA, A., 1965. Aspetti dell'evoluzione magmatica del vulcano di Roccamonfina attraverso misure di radioattività. *Ann. Osserv. Vesuviano*, 7: 77–105.

CIVETTA, L., GASPARINI, P. and ADAMS, J. A. S., 1969. ^{40}K determination by gamma-ray spectrometry for ^{40}K–^{40}Ar geochronology. *Anal. Chem.*, 41: 1319–1322.

CLARK, S. P., 1961. Heat flow from a differentiated earth. *J. Geophys. Res.*, 66: 1231–1234.

COCKROFT, A. L. and CURRAN, S. C., 1951. The elimination of the end effect in proportional counters. *Rev. Sci. Instr.*, 22: 37–42.

COHEN, J., ONCESCU, M. and REBIGAN, FL., 1964. Apparatus for relative measurements of gamma activities provided with 4π ionization chamber (well type). *Rev. Roumaine Phys.*, 9: 479–486.

COLLIE, C. H., HALBAN, H. and WILSON, R., 1950. Photoelectric dissociation of the Deuteron. *Proc. Phys. Soc. London, Ser. A*, 63: 994–1009.

CROUTHAMEL, C. E. (Editor), 1960. *Applied Gamma Ray Spectrometry*. Pergamon, London, 443 pp.

CROZAZ G., 1967. Datation des glaciers par le plomb-210. In: *Radioactive Dating and Methods of Low Level Counting*. International Atomic Energy Agency, Vienna, pp.385–393.

CROZAZ, G., PICCIOTTO, E. and DEBREUCK, W., 1964. Antarctic snow chronology with Pb-210. *J. Geophys. Res.*, 69: 2597–2604.

CULHANE, J. L., HERRING, J., SANFORD, P. W., O'SHEA, G. and PHILLIPS, R. D., 1966. Advances in the design and performance of X-ray proportional counters. *J. Sci. Instr.*, 43: 908–912.

CURIE, I. et FORAGGI, H., 1951. Autoradiographie par neutrons. Dosage separé de l'uranium et du thorium dans les mineraux. *Compt. Rend.*, 232: 959–961.

CURRAN, S. C., 1953. *Luminescence and the Scintillation Counter*. Academic Press, New York, N.Y., 213 pp.

CURRAN, S. C., 1955. Proportional counters. In: K. SIEGBAHN (Editor), *Beta and Gamma Ray Spectroscopy*. North-Holland, Amsterdam, pp.165–183.

CURRAN, S. C., 1958. Proportional counters. In: S. FLÜGGE and E. CREUTY (Editors), *Encyclopedia of Physics*. Springer, Berlin, pp.174–221.

CURRAN, S. C. and CRAGGS, J. D., 1949. *Counting Tubes: Theory and Application*. Academic Press, New York, N.Y., 238 pp.

CURRAN, S. C. and WILSON, H. W., 1965. Proportional counters and pulse ion chambers. In: K. SIEGBAHN (Editor), *Alpha, Beta and Gamma Ray Spectroscopy*. North-Holland, Amsterdam, pp.303–344.

CURRAN, S. C., ANGUS, J. and COCKROFT, A. L., 1948. Beta spectrum of tritium. *Nature*, 162: 302–303.

CURRAN, S. C., COCKROFT, A. L. and ANGUS, J., 1949. Investigation of soft radiations by proportional counters. V. Use as a detector of ultraviolet quanta and analysis of the multiplication process. *Phil. Mag.*, 40: 929–937.

CURRAN, S. C., COCKROFT, A. L. and INSCH, G. M., 1950. Gamma and beta spectroscopy with proportional counters in magnetic fields. *Proc. Phys. Soc. London, Ser. A*, 63: 845–847.

CURRIE, L. A., 1968. Limits for qualitative detection and quantitative determination, application to radiochemistry. *Anal. Chem.*, 40: 586–593.

CZUBEK, J. A., 1965. Physical possibilities of gamma–gamma logging. In: *Radioisotopes Instruments and Geophysics*. International Atomic Energy Agency, Vienna, pp.35–52.

CZUBEK, J. A., 1968. Influence of borehole construction on the results of spectral gamma logging. *Symp. Use Nuclear Techniques Prospecting, Develop. Mineral Resources, Buenos Aires*. International Atomic Energy Agency, Vienna, 13 pp.

CZUBEK, J. A. and DUMESNIL, D., 1968. Selective natural gamma-logging. *Symp. Use Nuclear Techniques Prospecting, Develop. Mineral Resources, Buenos Aires*, International Atomic Energy Agency, Vienna, 10 pp.

CZUBEK, J. A. and LENDA, A., 1968. Energy distribution of scattered gamma-rays in natural gamma-logging. *Symp. Use Nuclear Techniques Prospecting, Develop. Mineral Resources, Buenos Aires*. International Atomic Energy Agency, Vienna, 11 pp.

DANCHENKO, V. and MITROFANOV, N., 1964. Chlorine–neon Geiger–Mueller counters with energy independent gamma-ray efficiency. *Rev. Sci. Instr.*, 35: 1076–1077.

DANIELS, F., 1954. Radioactivity, energy storage and volcanism. In: H. FAUL (Editor), *Nuclear Geology*. Wiley, New York, N.Y., pp.188–194.

DARNLEY, A. G., 1968. Helicopter tests with a gamma-ray spectrometer. *Can. Mining J.*, 104–106.

DARNLEY, A. G. and FLEET, M., 1968. Evaluation of air-borne gamma-ray spectrometry in the Bancroft and Ellitt Lake areas of Ontario, Canada. In: D. C. PARKER (Editor), *Proceedings, 5th Symposium on Remote Sensing of Environment*. University of Michigan, Ann Arbor, Mich., pp.833–853.

DARNLEY, A. G., BRISTOW, Q. and DONHOFFER, D. K., 1968. Air-borne gamma-ray spectrometer experiments over the Canadian Shield. *Symp. Use Nuclear Techniques Prospecting, Develop. Mineral Resources, Buenos Aires*, presented paper.

DAVISSON, C. M., 1965. Interaction of gamma radiation with matter. In: K. SIEGBAHN (Editor), *Alpha, Beta and Gamma Ray Spectroscopy, 1*. North-Holland, Amsterdam, pp.37–78.

DEAN, P. N., 1964. Computer techniques in gamma spectrometry. In: J. A. S. ADAMS and W. M. LOWDER (Editors), *The Natural Radiation Environment*. University of Chicago Press, Chicago, Ill., pp.617–626.

DEARNALEY, G. and NORTHRUP, D. C., 1966. *Semiconductor Counters for Nuclear Radiations*. Wiley, New York, N.Y. 331 pp.

DELIBRIAS, G. et REPAIRE, J. L., 1967. Etude du mouvement propre d'une installation a bas niveau. En: *Radioactive Dating and Methods of Low Level Counting*. International Atomic Energy Agency, Vienna, pp.603–611.

DENNEN, W. H., AHRENS, L. H. and FAIRBAIRN, H. W., 1951. Spectrochemical analysis of major constituent elements in mineral and rocks. *U.S. Geol. Surv. Bull.*, 980: 25–52.

DER MATEOSIAN, E. and GOLDHABER, R., 1966. Search for double decay in Ca-48. *C.E.R.N. Bull.*, 65–32: 35–40.

DEUTSCH, S., 1957. Ages des halos pleochroiques de la granodiorite de Lausitz. *Experientia*, 13: 309.

DEUTSCH, S., HIRSCHBERG, D. et PICCIOTTO, E., 1956. Étude quantitative des halos pleochroiques. Application à l'estimation de l'age des roches granitique. *Bull. Soc. Belge Géol. Paléontol. Hydrol.*, 65: 267–273.

DE VRIES, H. and BARENDSEN, G. W., 1953. Radiocarbon dating by a CO_2 filled proportional counter. *Physica*, 19: 987–1003.

DIETHORN, W., 1956. A methane proportional counter system for natural radiocarbon measurement. *N.Y.O. Rept.*, 6628.

DODD, P. H., 1966. Quantitative logging and interpretation systems to evaluate uranium deposits. *Trans. Soc. Profess. Well Logging Analysts*, 1–16.

DODD, P. H. and DROULLARD, R. F., 1965. A logging system and computer program to determine rock density of uranium deposits. In: *Radioisotopes Instruments and Geophysics*. International Atomic Energy Agency, Vienna.

DODD, P. H., DROULLARD, R. F. and LATHAN, C. P., 1967. Borehole logging methods for exploration and evaluation of uranium deposits. *Can. Centennial Conf. Mining Geophysics, Niagara*, presented paper.

DOIG, R., 1968. The natural gamma-ray flux: in situ analysis. *Geophysics*, 33: 311–328.

DOKE, T., 1962. A new method for the reduction of alpha ray background in a gridded ionization chamber. *Can. J. Phys.*, 40: 607–621.

DOMINGOS, J. M. and MELO, A. A., 1967. Determination of uranium and thorium ores using gamma spectrometry. *Nucl. Instr. Methods*, 47: 305–319.

DONHOFFER, D. K., 1968. Determination of ore concentration exploration holes by the gamma-ray backscattering technique. *Symp. Use Nuclear Techniques*

Prospecting, Develop. Mineral Resources, Buenos Aires. International Atomic Energy Agency, Vienna, 11 pp.

DOOLEY, J. R., TATSUMOTO, M. and ROSHOLT, J. N., 1964. Radioactive disequilibrium studies of roll features, Shirley Basin, Wyoming. *Econ. Geol.*, 59: 586–595.

DRAGHICESCU, M., 1961. Mise au point des compteurs Geiger–Mueller functionnant à haute temperature. *Nucl. Electronics*, 1: 345–352.

DREVER, R. P. W., MOLJK, A. and CURRAN, S. C., 1957. A proportional counter with small wall effect. *Nucl. Instr.*, 1: 41–45.

DUIVENSTIJN, A. J. and VENVERLOO, L. A. J., 1963. *Practical Gamma Spectrometry.* Charles Thomas, Springfield, Ill., 156 pp.

EATON, J. P. and MURATA, K. J., 1960. How volcanoes grow. *Science*, 132: 925–938.

EBERHART, P., GEISS, J., HOUTERMANS, F. G., BUSER, W. e VON GUNTEN, H. R., 1955. Il piombo vulcanico del Vesuvio e di Vulcano. *Atti Conv. Geol. Nucl., 1, Rome, 1955*, pp.15–20.

EBERT, K. H., HERNEGGER, F., KONIG, H. und WANKE, H., 1959. Der Meteorit von Breitscheid. IV.C. Uran. *Geochim. Cosmochim. Acta*, 17: 349–350.

EBERT, K. H., KONIG, H. und WANKE, H., 1957. Eine neue Methode zur Bestimmung kleinsten Uranmengen und ihre Anwendung auf die Urananalyse von Steinmeteoriten. *Z. Naturforsch.*, 12a: 763–765.

ECHO, M. W. and TURK, E. K., 1957. Quantitative determination of ^{235}U by delayed neutrons counting. *U.S. At. Energy Comm., Rept.*, PTR 143: 80 pp.

ECKELMANN, W. R., BROECKER, W. S. and KULP, J. L., 1960. Half life of Pb-210. *Phys. Rev.*, 118: 668–671.

EDERLYVARI, I. and FEHER, I., 1967. 4π ionization chamber for measurement of low energy gamma emitters. *Nucl. Instr. Methods*, 54: 163–164.

EGELKRAUT, K. und LEUTZ, H., 1960. Halbwertszeit des ^{40}K. *Phys. Verhandl.*, 11: 67–80.

EICHHOLZ, G. G., HILBORN, J. W. and MACMAHON, C., 1953. The determination of uranium and thorium in ores. *Can. J. Phys.*, 31: 613–628.

EL SHISHINI, M. M. and ZOBEL, W., 1966. Temperature dependence of the response of lithium-drifted germanium detectors to gamma-rays. *Inst. Elec. Electron. Engrs., Trans. Nucl. Sci.*, 13: 359–362.

EVANS, R. D., 1955. *The Atomic Nucleus.* McGraw-Hill, New York, N.Y., 1120 pp.

FACCHINI, U. e ORSONI, L., 1949. Un metodo utilizzante la scissione dell' ^{235}U per l'analisi del contenuto in uranio nei minerali. *Nuovo Cimento*, 6: 241–254.

FANO, U., 1947. Ionization yields of radiations. Fluctuations of the number of ions. *Phys. Rev.*, 72: 26–29.

FANO, U., BERGER, C. D. and ZERBER, C. D., 1962. Gamma ray attenuation. In: E. P. BLIZARD and L. S. ABBOTT (Editors), *Reactor Handbook, 3, Part B.* Interscience, New York, N.Y., 2nd ed., pp.102–127.

FAUL, H. (Editor), 1954. *Nuclear Geology.* Wiley, New York, N.Y., 415 pp.

FAUL, H. and TITTLE, C. W., 1951. Logging of drill holes by the neutron–gamma method and gamma–gamma scattering. *Geophysics*, 16: 261–276.

FAUST, W. R., 1950. Specific activity of potassium. *Rev. Phys.*, 78: 624.

FERRETTI SFORZINI, M., FESTA, C. e IPPOLITO, F., 1955. Sulla dipendenza dalla data di eruzione della radioattività delle lave vesuviane. *Atti Conv. Geol. Nucl., 1, Rome, 1955*, pp.32–38.

FINK, R. W. and ROBINSON, B. L., 1956. Compton backscattered photons in precision proportional counter spectrometry. *Rev. Sci. Instr.*, 27: 712–713.

FLANAGAN, F. J., 1967. U.S. Geological Survey silicate rock standards. *Geochim. Cosmochim. Acta*, 31: 289–308.

FLANAGAN, F. J., 1969. U.S. Geological Survey rock standards. II. First compilation of data for the new U.S.G.S. rocks. *Geochim. Cosmochim. Acta*, 33: 81–120.

FLEISCHER, M., 1965. Summary of new data on rock samples G-1 and W-1: 1962–1965. *Geochim. Cosmochim. Acta*, 29: 1263–1283.

FLEISCHER, M., 1969. U.S. Geological Survey standards: I. Additional data on rocks G-1 and W-1, 1965–1967. *Geochim. Cosmochim. Acta*, 33: 65–80.

FLEISCHER, M. and STEVENS, R. E., 1962. Summary of new data on rock samples G-1 and W-1. *Geochim. Cosmochim. Acta*, 26: 525–543.

FLEISCHER, R. L. and PRICE, P. B., 1964a. Fission track evidence for simultaneous origin of tektites and other natural glasses. *Geochim. Cosmochim. Acta*, 28: 755–760.

FLEISCHER, R. L. and PRICE, P. B., 1964b. Techniques for geological dating of minerals by chemical etching of fission fragments tracks. *Geochim. Cosmochim. Acta*, 28: 1705–1714.

FLEISCHER, R. L., PRICE, P. B. and WALKER, R. M., 1965a. Solid state track detectors: applications to nuclear science and geophysics. *Ann. Rev. Nucl. Sci.*, 15: 1–28.

FLEISCHER, R. L., NAESER, C. W., PRICE, P. B., WALKER, R. M. and MARVIN, U. B., 1965b. Fossil particle tracks and uranium distribution in minerals of the Vaca Muerta meteorite. *Science*, 148: 629–632.

FLEMING JR, E. H., GHIORSO, A. and CUNNINGHAM, B. B., 1952. The specific alpha activities and half lives of ^{234}U, ^{235}U, and ^{236}U. *Phys. Rev.*, 88: 642–652.

FOOTE, R. S., 1968a. Application of air-borne gamma radiation measurements to pedologic mapping. In: D. C. PARKER (Editor), *Proceedings, 5th Symposium on Remote Sensing of the Environment*. University of Michigan, Ann Arbor, Mich.

FOOTE, R. S., 1968b. Improvements in air-borne gamma-radiation data analyses for anomalous radiation by removal of environmental and pedologic radiation changes. *Symp. Use Nuclear Techniques Prospecting, Develop. Mineral Resources, Buenos Aires*. International Atomic Energy Agency, Vienna, 8 pp.

FORNACA-RINALDI, G., 1968a. Il metodo $^{230}Th/^{238}U$ per la datazione di stalattiti e stalagmiti. *Boll. Geofis. Teor. Appl.*, 10: 3–14.

FORNACA-RINALDI, G., 1968b. $^{230}Th/^{234}Th$ dating of cave concretions. *Earth Planetary Sci. Letters*, 5: 120–122.

FRANK, M., DUBRAU, H. J., LINGERTAT, H. and HERFORT, L., 1964. Thermoluminescent dosimetry with lithium fluoride. *Kernenergie*, 7: 570–573.

GALE, N. H., 1967. Development of delayed neutrons technique as a rapid and precise method for determination of uranium and thorium at trace levels in rocks and minerals, with applications to isotope geochronology. In: *Radioactive Dating and Methods of Low Level Counting*. International Atomic Energy Agency, Vienna, pp.431–452.

GASPARINI, P., 1963. Radioactivity of the lavas of the island of Ischia. *Ann. Osserv. Vesuviano*, 5: 185–218.

GASPARINI, P. and ADAMS, J. A. S., 1969. K–Ar dating of Italian Plio-Pleistocene volcanoes. *Earth Planetary Sci. Letters*, 6: 225–230.

GASPARINI, P. and LUONGO, G., 1965. Radioattività artificiale alla Solfatara di Pozzuoli nella primavera del 1965. *Boll. Soc. Nat., Napoli*, 74: 205–216.

GASPARINI, P., RAPOLLA, A. e SABINI, P., 1961. Considerazioni su di un rilevamento radiometrico dell'isola d'Ischia. *Atti Conv. Assoc. Geofis. Ital., 11, Rome*, pp.195–208.

GASPARINI, P., LUONGO, G. e DAVIA, G., 1963. Misure di radioattività alla Solfatara di Pozzuoli. *Atti Conv. Assoc. Geofis. Ital., 12, Rome*, pp.41–50.

GASPARINI, P., LIRER, L. and LUONGO, G., 1967. Petrochemical and physical characteristics of the lava erupted from Stromboli in April 1967. *Ann. Osserv. Vesuviano*, 8: 15–32.

GAST, P. W., 1960. Limitations on the composition of the upper mantle. *J. Geophys. Res.*, 65: 1287–1297.

GATROUSIS, C., HEINRICH, R. and CROUTHAMEL, C. E., 1961. Recent advances in counting techniques. In: *Analytical Chemistry, S.IX, 2.* Pergamon, New York. N.Y., pp.1–79.

GIBSON, W. M., MILLER, G. L. and DANOVAN, P. F., 1965. Semiconductor particle spectrometers. In: K. SIEGBAHN (Editor), *Alpha, Beta and Gamma Ray Spectroscopy, 1.* North-Holland, Amsterdam, pp.345–378.

GIMENEZ, G. et LABEYRIE, J., 1952. Fabrication et propriétés des compteurs de Geiger remplis à la vapeur de Brome et destinés à la prospection beta et gamma. *Nuovo Cimento*, 9: 169–183.

GIRARDI, F., 1961. Development of new techniques of activation analysis at the Community Research Center of Euratom, Ispra. *Proc. Intern. Conf. Modern Trends Activation Analysis*, pp.119–125.

GIRDLER, R. W., 1967. A review of terrestrial heat flow. In: S. K. RUNCORN (Editor), *Mantles of the Earth and Terrestrial Planets.* Interscience, London, pp.549–566.

GLASSTONE, S., 1958. *Sourcebook on Atomic Energy*. Van Nostrand, Princeton, N.J., 641 pp.

GLENDENIN, L. E., 1961. The present state of the decay constants. *Ann. N.Y. Acad. Sci.*, 91: 166–181.

GLOS, M. B., 1964. Semiconductors, scintillators and data analysis. *Nucleonics,* 22: 50–53.

GOGANOV, D. A. and GOGOLEV, G. P., 1963. Proportional counters for detecting X-radiation. *Izv. Akad. Nauk S.S.S.R., Ser. Phys.*, 27: 438–445.

GOLDBERG, E. D., 1963. Geochronology with Pb-210. In: *Radioactive Dating and Methods of Low Level Counting*. International Atomic Energy Agency, Vienna, pp.121–130.

GOLDSWORTHY, D., 1960. Transistorized portable counting rate meter. *Nucleonics*, 18: 92–99.

GORDON, G. E., RANDLE, K., GOLES, C. G., CORLISS, J. B., BEESON, M. H. and OXLEY, S., 1968. Instrumental activation analysis of standard rocks with high resolution gamma-ray detectors. *Geochim. Cosmochim. Acta*, 32: 369–398.

GOULDING, F. S., 1964. Semiconductor detectors: their properties and applications. *Nucleonics*, 22: 54–61.

GOULDING, F. S., 1965. Multichannel pulse amplitude analyzers. In: K. SIEGBAHN (Editor), *Alpha, Beta and Gamma Ray Spectroscopy, 1.* North-Holland, Amsterdam, pp.413–424.

GOULDING, F. S., 1966. Semiconductor detectors for nuclear spectrometry. *Nucl. Instr. Methods.*, 43: 1–54.

GRAEFFE, G. and NURMIA, M., 1961. The use of thick sources in alpha spectrometry. *Ann. Acad. Sci. Fennicae, Ser. A*, 6: 1–14.

GREEN, D. H., 1968. Origin of basaltic magmas. In: H. H. HESS and A. POLDERVAART (Editors), *The Poldervaart Treatise on Rocks of Basaltic Composition, 2.* Interscience, New York, N.Y., pp.835–862.

GREEN, D. H., MORGAN, J. W. and HEIER, K. G., 1968. Thorium, uranium and potassium abundances in periodotite inclusions and their host basalts. *Earth Planetary Sci. Letters*, 4: 155–166.

GRIMALDI, F. S., 1961. Thorium. In: I. M. KOLTHOFF and P. J. ELVING (Editors),

Treatise on Analytical Chemistry, Part II, 5. Interscience, New York, N.Y., pp.139–216.

GRIMALDI, F. S., MAY, I., FLETHCHER, M. H. and TITCOMB, J., 1954. Collected papers on methods of analysis for uranium and thorium. *U.S. Geol. Surv. Bull.*, 1006: 1–184.

GRINBERG, B. et LE GALLIC, J., 1961. Caractéristiques fondamentales d'un laboratoire de mesure de tres faible activités. *Intern. J. Appl. Radiation Isotopes*, 12: 104–117.

GRINDLER, J. E., 1962. The radiochemistry of uranium. *Natl. Acad. Sci.—Natl. Res. Council, Nucl. Sci. Ser.*, 3050.

GRISSOM, J. T. and KOEHLER, D. R., 1965. Single crystal Compton spectroscopic techniques. *Nucl. Instr. Methods*, 37: 336–339.

GRODSTEIN, W. R., 1957. X-ray attenuation coefficients from 10 keV to 100 Mev. *Natl. Bur. Std., (U.S.) Circ.*, 583.

GRUMMITT, W. E., BROWN, R. M., CRUIKSHANK, A. J. and FOWLER, I. L., 1956. Recent developments in low background Geiger–Mueller counters. *Can. J. Chem.*, 34: 206–213.

GUILBERT, J. and ADAMS, J. A. S., 1955. Alpha particle autoradiography with nuclear emulsions. *Nucleonics*, 13: 43.

GUILLOU, R. B., 1964. The aerial radiological measuring surveys (ARMS) program. In: J. A. S. ADAMS and W. M. LOWDER (Editors), *The Natural Radiation Environment.* University of Chicago Press, Chicago, Ill., pp.705–722.

HABERER, K., 1965. Über den Wirkungsgrad von gamma-Aktivitätmessungen von Scintillationszählern. *Fresenius' Z. Anal. Chem.*, 210: 412–418.

HAMAGUCHI, H., REED, G. W. and TURKEVICH, A., 1957. Uranium and barium in stone meteorites. *Geochim. Cosmochim. Acta*, 12: 337–347.

HAMILTON, E. I., 1959. The uranium content of the differentiated Skaergaard intrusion. *Medd. Grønland*, 162: 4–34.

HAMILTON, E. I., 1966. The determination of uranium in rocks and minerals by the delayed neutrons method. *Earth Planetary Sci. Letters*, 1: 77–81.

HAMILTON, E. I., 1968. The isotopic composition of strontium applied to problems of the origin of alkaline rocks. In: E. I. HAMILTON and R. M. FARQUHAR (Editors), *Radiometric Dating for Geologists.* Interscience, London, pp.437–464.

HAND, J. A., 1964. Instrumentation for aerial surveys of terrestrial gamma radiation. In: J. A. S. ADAMS and W. M. LOWDER (Editors), *The Natural Radiation Environment.* University of Chicago Press, Chicago, Ill., pp.687–704.

HANSEN, P. G., NIELSEN, H. L., WILLIAMS, E. T. and WILSKY, K., 1965. Low energy gamma spectroscopy using an X-ray escape gated proportional counter. *Nucl. Instr. Methods*, 32: 197–200.

HARBOTTLE, G., 1959. Half lives of ^{204}Tl, RaD and ^{207}Bi. *J. Inorg. Nucl. Chem.*, 12: 6–7.

HARRIS, P. G., 1967. Segregation processes in the upper mantle. In: S. K. RUNCORN (Editor), *Mantles of the Earth and Terrestrial Planets.* Interscience, London, pp.305–317.

HART, R. J., RUSSEL, K. and STEFFEN, R. M., 1951. Quantum efficiency of Bismuth counters for electromagnetic radiations of 0.07 Mev to 2.8 Mev energies. *Phys. Rev.*, 81: 460–461.

HAWKINS, W. K. and GEARHART, M., 1968. Gamma-ray logging in uranium prospecting. *Symp. Use Nuclear Techniques Prospecting, Develop. Mineral Resources, Buenos Aires.* International Atomic Energy Agency, Vienna, 10 pp.

HEATH, R. L., 1957. Scintillation spectrometry. Gamma ray spectrum catalogue. *U.S. At. Energy Comm., Rept.*, IDO 16408.

HEATH, R. L., 1961. The development of a spectrum catalogue and data processing techniques for the gamma-ray scintillation spectrometer. *Proc. Intern. Conf. Modern Trends Activation Analysis*, pp.155–160.

HEATH, R. L., 1966. Computer techniques for the analysis of gamma-ray spectra obtained with NaI(Tl) and lithium ion drifted Germanium detectors. *Nucl. Instr. Methods*, 43: 209–229.

HEIER, K. S. and ADAMS, J. A. S., 1965. Concentration of radioactive elements in deep crustal materials. *Geochim. Cosmochim. Acta*, 29: 53–62.

HEIER, K. S. and CARTER, J. L., 1964. Uranium, thorium and potassium content in basic rocks and their bearing on the nature of the upper mantle. In: J. A. S. ADAMS and W. M. LOWDER (Editors), *The Natural Radiation Environment*. University of Chicago Press, Chicago, Ill., pp.63–86.

HEIER, K. S. and ROGERS, J. J. W., 1963. Radiometric determinations of thorium, uranium and potassium in basalts and in two magmatic differentiation series. *Geochim. Cosmochim. Acta*, 27: 137–154.

HELMER, R. G., HEATH, R. L., SCHMITTROTH, L. A., JAYNE, G. A. and WAGNER, L. M., 1967a. Analysis of gamma ray spectra from Ge(Li) and NaI(Tl) spectrometers. Computer program. *Nucl. Instr. Methods*, 47: 305–319.

HELMER, R. G., HEATH, R. L., PUTNAM, M. and GIPSON, D. H., 1967b. Photopeak analysis program for photon energy and intensity determination. NaI(Tl) and Ge(Li) spectrometers. *Nucl. Instr. Methods*, 57: 46–57.

HENGELCKE, B. A. und HÜBNER, W., 1965. Untersuchungen an Hohlraum-Ionizationskammern zur Messung der Ionendosis bei Röntgen und Gamma-Strahlungen im Photonenenergie zwischen 100 kev and 1.3 Mev. *Biophysik*, 2: 193–211.

HERPERS, U., HERR, W. and WOLFE, R., 1967. Determination of cosmic ray produced nuclides ^{53}Mn, ^{45}Sc and ^{26}Al in meteorites by neutron activation and gamma coincidence spectrometry. In: *Radioactive Dating and Methods of Low Level Counting*. International Atomic Energy Agency, Vienna, pp.199–206.

HILL, C. R., OSBORNE, R. V. and MAYNEFORD, W. V., 1964. Studies of alpha radioactivity in relation to man. In: J. A. S. ADAMS and W. M. LOWDER (Editors), *The Natural Radiation Environment*. University of Chicago Press, Chicago, Ill., pp.395–406.

HOFSTADTER, R., O'DELL, E. W. and SCHMIDT, C. D., 1964. CaI_2 and CaI_2(Eu) scintillation crystals. *Inst. Elec. Electron. Engrs., Trans. Nucl. Sci.*, 11: 12–15.

HOLLANDER, J. M., 1966. The impact of semiconductor detectors on gamma ray and electron spectroscopy. *Nucl. Instr. Methods*, 43: 65–109.

HOMILIUS, J. and LORCH, S., 1957. On the theory of gamma-ray scattering in boreholes. *Geophys. Prospecting*, 5: 449–468.

HOUTERMANS, F. G., 1961: Thermoluminescence glow curves. In: F. G. HOUTERMANS, E. PICCIOTTO and E. TONGIORGI (Editors), *Summer Course in Nuclear Geology*. Laboratorio di Geologia Nucleare, Pisa, Italy, pp.233–240.

HOUTERMANS, F. G., 1966. History of the K/Ar method. In: O. A. SCHAEFFER and J. ZÄHRINGER (Editors), *Potassium–Argon Dating*. Springer, New York, N.Y., pp.1–6.

HOUTERMANS, F. G., HAXEL, O. und HEINTZ, J., 1950. Die Halbwertszeit des K-40. *Z. Physik*, 128: 657–667.

HOUTERMANS, F. G., EBERHART, P. and FERRARA, G., 1964. Lead of volcanic origin. In: H. CRAIG, S. L. MILLER and G. J. WASSERBURG (Editors), *Isotopic and Cosmic Chemistry*. North-Holland, Amsterdam, pp. 233–244.

HUANG, W. H., MAURETTE, M. and WALKER, R. M., 1967. Observations of fossil alpha particles recoil tracks and their implications for dating measurements. In:

Radioactive Dating and Methods of Low Level Counting. International Atomic Energy Agency, Vienna, pp.415–430.

Hugi, T. et Jedwab, J., 1966. Distribution de la radioactivité et phénomènes optiques dans le charbon de Dorenaz (Canton du Valais, Suisse). *Schweiz. Mineral. Petrog. Mitt.*, 46: 377–387.

Hurley, P. M., 1954. Use of the gamma ray scintillation spectrometer in the separate measurement of the U and Th series in geological materials. *Bull. Geol. Soc. Am.*, 65: 1265–1266 (Abstract).

Hurley, P. M., 1956. Direct radiometric measurement by gamma-ray scintillation spectrometer. Part I, II. *Bull. Geol. Soc. Am.*, 67: 395–420.

Hurley, P. M., Fairbairn, H. W. and Pinson Jr., W. H., 1966. Rb–Sr isotopic evidence in the origin of potash rich lavas of western Italy. *Earth Planetary Sci. Letters*, 1: 301–306.

Hyde, E. K., 1964. *The Nuclear Properties of the Heavy Elements*. III. *Fission Phenomena*. Prentice Hall, Englewood Cliffs, N.J., 519 pp.

Hyde, E. K., Perlman, I. and Seaborg, G. T., 1964. *The Nuclear Properties of the Heavy Elements. II. Detailed Radioactive Properties*. Prentice Hall, Englewood Cliffs, N.J., 1107 pp.

Hyndmann, R. D., Lambert, I. B., Heier, K. S., Jaeger, J. C. and Ringwood, A. E., 1968. Heat flow and surface radioactivity measurements in the Pre-cambrian shield of western Australia. *Phys. Earth Planetary Interior*, 1: 129–135.

Imbo', G., 1939. Ulteriori misure di radioattività delle acque di Lacco Ameno e Cetara. *Ric. Sci.*, 10: 15–37.

Imbo', G., 1955. Studio radiometrico del Gran Cono Vesuviano. *Atti Conv. Geol. Nucl. 1, Rome* (abstract), 27 pp.

Imbo', G., Vittozzi, P. e Gasparini, P., 1965. Possibilità di datazione di lave vesuviane mediante la radioattività. *Atti Accad. Pontaniana Napoli*, 14:215–223.

Imbo', G., Gasparini, P., Luongo, G. and Rapolla, A., 1968. Contributions to the volcanological researches by determination of the radioactivity of eruptive products. *Bull. Volcanol.*, 32, in press.

Inghram, M. G., Brown, H., Patterson, C. and Hess, D. C., 1950. The branching ratio of K-40 radioactive decay. *Phys. Rev.*, 80: 916–917.

Israël, H., 1964a. The radon-220 content of the atmosphere. In: J. A. S. Adams and W. M. Lowder (Editors), *The Natural Radiation Environment*. University of Chicago Press, Chicago, Ill., pp.313–314.

Israël, H., 1964b. Remarks on the thoron content of the atmosphere. *J. Atmospheric Terrest. Phys.*, 26: 787–789.

Israël, H. and Israël, G. W., 1965. Measurements of the thoron content of the atmosphere and their application in meteorology. *Final Tech. Rept.*, DA 91–501, EUC- 3483.

Jacobs, A. W. B., 1964. Spherical Geiger counter for gamma-ray flux measurement. *Intern. J. Appl. Radiation Isotopes*, 15: 583.

Jacobs, A. W. B., 1965. Design and use of spherical Geiger counters. *Sci. Proc. Roy. Dublin Soc., Ser. A*, 2: 43–46.

Jarczyk, L., Kroepfel, H., Lang, J., Müller, R. and Wolff, W., 1962. Photo-peak efficiency and response function of various NaI(Tl) and CsI(Tl) crystals in the energy range up to 11 Mev. *Nucl. Instr. Methods*, 17: 310–320.

Jedwab, J., 1965. Les dégâts radiatifs dans le charbon uranifère du Schaentzel. *Geol. Rundschau*, 55: 445–453.

Joly, J., 1909. On the radioactivity of certain lavas. *Phil. Mag.*, 18: 115–118.

JURNEY, E. T. and MAIENSCHEIN, F., 1949. The gamma ray counting efficiency for a lead cathode GM counter. *Rev. Sci. Instr.*, 20: 932–934.

KAISER, H., 1965. Zum Problem der Nachweisgrenze. *Fresenuis' Z. Anal. Chem.*, 209: 1–18.

KAISER, W. G., 1966. Scintillation spectrometry: the state of the art. *Anal. Chem.*, 38: 27A–40A.

KAISER, W. and ZÄHRINGER, J., 1968. K/Ar age determinations of iron meteorites. IV. New results with refined experimental procedures. *Earth Planetary Sci. Letters*, 4: 84–88.

KAPLAN, I., 1963. *Nuclear Physics*. Addison–Wesley, Reading, Mass., 770 pp.

KAUFHOLD, J. and HERR, W., 1967. Influence of experimental factors on dating natural and man-made glasses by fission-tracks method. In: *Radioactive Dating and Methods of Low Level Counting*. International Atomic Energy Agency, Vienna, pp.403–411.

KELLY, G. G., BELL, P. R., DAVIS, R. C. and LAZAR, N. H., 1956. Intrinsic scintillation resolution. *Inst. Radio Engrs., Trans. Nucl. Sci.*, 3(4): 57–58.

KELLY, W. H., BEARD, G. B. and PETERS, R. A., 1959. The beta decay of ^{40}K. *Nucl. Phys.*, 11: 492–498.

KENDALL, B. R. F., 1960. Isotopic composition of potassium. *Nature*, 186: 225–226.

KHRISTOV, L., 1957. On the dependence of the gas amplification factor with the proportional counters voltage. *Dokl. Akad. Nauk Bulgar.*, 10: 453–456.

KIGOSHI, K., 1967. Ionium dating of igneous rocks. *Science*, 156: 932–934.

KING, J. D., NEFF, N. and TAYLOR, H. W., 1967. The energy of ^{40}K gamma ray and its use as a calibration standard. *Nucl. Instr. Methods*, 52: 249–250.

KIRKWOOD, D. H. W., PONTECORVO, B. and HANNA, G. C., 1948. Fluctuations of ionization and low energy gamma spectra. *Phys. Rev.*, 74: 497–498.

KLEEMAN, J. D. and LOVERING, J. F., 1967. Uranium distribution in rocks by fission tracks registration in Lexan Plastic. *Science*, 156: 512–513.

KOCHAROV, G. E., and KOROLEV, G. A., 1963. Theory and operation of proportional counters. *Izv. Akad. Nauk S.S.S.R., Ser. Phys.*, 27: 301–308.

KOCZY, F. F., PICCIOTTO, E., POULAERT, G. et WILGAIN, S., 1957. Mesure des isotopes du thorium dans l'eau de mer. *Geochim. Cosmochim. Acta*, 11: 103–129.

KOEHLER, D. R. and GRISSOM, J. T., 1967. New X-ray spectrometric technique. *Nucl. Instr. Methods*, 50: 87–90.

KOHMAN, T. P., 1956. Extinct natural radioactivity. *Ann. N.Y. Acad. Sci.*, 62: 503–542.

KOHMAN, T. P., 1961. Chronology of nucleosynthesis and extinct natural radioactivity. *J. Chem. Educ.*, 38: 73–82.

KÖNIG, H. und WANKE, H., 1959. Uranbestimmungen an Steinmeteoriten mittels Neutronaktivierung über die Xenon-isotope 133 und 135. *Z. Naturforsch.*, 14a: 866–869.

KORKISCH, J. and JANAUER, G. E., 1962. Ion exchange in mixed solvents: adsorption behaviour of uranium and thorium on strong base anion exchange resins from mineral acid–alcohol media: separation methods for uranium and thorium. *Talanta*, 9: 957–985.

KOWALSKI, B. R. and ISENHOUR, T. L., 1968. An analytical function for describing gamma ray pulse height distributions in NaI(Tl) scintillators. *Anal. Chem.*, 40: 1186–1193.

KOWNACKI, I. and MARCINKOWSKI, A., 1965. Proportional counter as a spectrometer of low energy beta and gamma rays. *Nukleonika*, 10: 669–673.

KRAUSKOPF, K., 1967. *Introduction to Geochemistry*. McGraw-Hill, New York, N.Y., 721 pp.

LACHENBRUCH, A. H., 1968. Preliminary geothermal model of the Sierra Nevada. *J. Geophys. Res.*, 73: 6977–6989.

LACHENBRUCH, H. A., WOLLENBERG, H. A., GREENE, G. W. and SMITH, A. R., 1966. Heat flow and heat production in the central Sierra Nevada, preliminary results. *Trans. Am. Geophys. Union*, 47: 179 (Abstract).

LAL, D. and PETERS, B., 1967. Cosmic ray produced activity in the earth. In: S. FLÜGGE (Editor), *Handbuch der Physik*. Springer, Berlin, pp.551–608.

LALOVIC, B., AZUMA, R. E. and PETROVIC, B., 1967. Large multiple element Germanium spectrometers. *Inst. Elec. Electron. Engrs., Trans. Nucl. Sci.*, 14: 514–520.

LAMBERT, I. B. and HEIER, K. S., 1967. The vertical distribution of uranium, thorium and potassium in the continental crust. *Geochim. Cosmochim. Acta*, 31: 377–390.

LEDERER, C. M., HOLLANDER, J. M. and PERLMAN, I., 1967. *Tables of Isotopes*, 4th ed. Wiley, New York, N.Y., 594 pp.

LEE, W. K. (Editor), 1965. Terrestrial heat flow. *Am. Geophys. Union, Publ.*, 1288: 276 pp.

LEGER, G., BIBRON, R., LABEYRIE, J. et DELIBRIAS, G., 1967. Spectromètre à grande sensibilité destiné à la mesure du rayonnement X de faible energie. In: *Radioactive Dating and Methods of Low Level Counting*. International Atomic Energy Agency, Vienna, pp.629–638.

LEIPUNSKII, O. I., NOVOZHILOV, B. V. and SAKHAROV, B. N., 1965. *The Propagation of Gamma Quanta in Matter*. Pergamon, New York, N.Y., 222 pp.

LESINSKI, J., 1965. Background stability of halogen quenched GM counters. *Nukleonika*, 10: 135–144.

LEUTZ, H., SCHULTZ, G. and WENNINGER, H., 1965. Decay of potassium 40. Z. *Physik*, 187: 151–172.

LEVINE, H. and GRIMALDI, F. S., 1958. Determination of thorium in the parts per million range in rocks. *Geochim. Cosmochim. Acta*, 14: 93–97.

LEVINE, H. and LAMANNE, H., 1965. Radiochemical determination of uranium in environmental media by electrodeposition. *U.S. Dept. Health Serv., Publ.*, 999-RH-1.

LEVY, A. J., RITTER, R. C. and ZIOCK, K., 1967. The annular lithium-drifted germanium detector as a standard geometry gamma spectrometer. *Inst. Elec. Electron. Engrs. Trans. Nucl. Sci.*, 14: 514–520.

LEVY, P. W., 1966. A brief survey of radiation effects applicable to geological problems. *Conf. Thermoluminescence Appl. Geological Problems, Spoleto, 1966*, 7 pp., presented paper.

LIPSCHUTZ, M. E., SIGNER, P. and ANDERS, E., 1965. Cosmic ray exposure ages of iron meteorites by the $^{21}Ne/^{26}Al$ method. *J. Geophys. Res.*, 70: 1473–1489.

LOCARDI, E. e SIRCANA, S., 1967. Distribuzione di Th, U e K nelle vulcaniti alcaline del Lazio settentrionale. *Boll. Soc. Mineral. Ital.*, 23: 163–224.

LOEVINGER, R. and BERMAN, M., 1951. Efficiency criteria in radioactivity counting. *Nucleonics*, 9: 26–39.

LONATI, R. D., FACCHINI, U., ZORI, Z., HOUTERMANS, F. G. and TONGIORGI, E., 1958. Study of alpha radioactivity in low concentrations. *Nuovo Cimento*, 7: 133–141.

LOOSLI, H. H., OESCHGER, H. and WAHLEN, M., 1967. New attempts in low level counting and a search for cosmic ray produced ^{36}Ar. In: *Radioactive Dating and*

Methods of Low Level Counting. International Atomic Energy Agency, Vienna, pp.593–602.

LÖVBÖRG, L., 1967a. A scaler-timer for portable instruments. *Nucl. Instr. Methods*, 54: 137–138.

LÖVBÖRG, L., 1967b. A portable gamma-ray spectrometer for field use. *Danish At. Energy Comm., Risö Rept.*, 168.

LÖVBÖRG, L., KUZENDORF, H. and HANSEN, J., 1968. Use of field gamma-spectrometry in the exploration of uranium and thorium deposits in South Greenland. *Symp. Use Nuclear Techniques Prospecting, Develop. Mineral Resources, Buenos Aires*. International Atomic Energy Agency, Vienna, 16 pp.

LOWDER, W. M., CONDON, W. J. and BECK, H. J., 1964. Field spectrometric investigation of environmental radioactivity in U.S.A. In: J. A. S. ADAMS and W. M. LOWDER (Editors), *The Natural Radiation Environment*. University of Chicago Press, Chicago, Ill., pp.597–616.

LUKIRSKII, A. P., BRYTOV, I. A. and ERSHOV, O. A., 1963. Proportional counter for soft X-rays. *Izv. Akad. Nauk S.S.S.R., Ser. Phys.*, 27: 446–451.

LUONGO, G. e RAPOLLA, A., 1964. Contributo allo studio dell'evoluzione del magma Somma–Vesuviano mediante le determinazione delle concentrazioni in isotopi radioattivi ^{238}U, ^{232}Th, ^{226}Ra, e ^{40}K. *Ann. Osserv. Vesuviano*, 6: 45–66.

LUONGO, G. e RAPOLLA, A., 1965. Spettrometria gamma delle lave della regione vulcanica dei Colli Albani (Vulcano Laziale). *Ann. Osserv. Vesuviano*, 7: 58–76.

MACDONALD, G. J. F., 1959. Calculations on the thermal history of the earth. *J. Geophys. Res.*, 64: 1967–2000.

MACDONALD, G. J. F., 1961. Surface heat flow from a differentiated earth. *J. Geophys. Res.*, 66: 2489–2493.

MACDONALD, G. J. F., 1963. The deep structure of continents. *Rev. Geophys.*, 1: 587–665.

MACDONALD, G. J. F., 1965. Geophysical deductions from observations of heat flow. In: H. K. LEE (Editor), *Terrestrial Heat Flow. Am. Geophys. Union, Publ.*, 1288: 191–210.

MACFARLANE, R. D. and KOHMAN, T. P., 1961. Natural alpha radioactivity in medium heavy elements. *Phys. Rev.*, 121: 1758–1769.

MAHLMAN, H. A. and LEDDICOTTE, G. W., 1955. Determination of microgram and submicrogram quantities of uranium by neutron activation analysis. *Anal. Chem.*, 27: 823–825.

MALM, H. L., 1966. Improvements in large volume coaxial germanium spectrometers. *Inst. Elec. Electron. Engrs., Trans. Nucl. Sci.*, 13: 285–289.

MALM, H. L., 1967. Encapsulated coaxial Ge(Li) detectors operating in parallel. *Inst. Elec. Electron. Engrs., Trans. Nucl. Sci.*, 14: 251–252.

MALM, H. L. and FOWLER, I. L., 1966. Large volume coaxial germanium gamma-ray spectrometers. *Inst. Elec. Electron. Engrs., Trans. Nucl. Sci.*, 13: 62–71.

MANAGAN, W. and CROUTHAMEL, C. E., 1960. Extrinsic variables. In: C. E. CROUTHAMEL (Editor), *Applied Gamma Ray Spectrometry*. Pergamon, New York, N.Y., pp.25–93.

MANDUCHI, C. and ZANNONI, G., 1962. A proportional counter with reduced wall effect. *Nucl. Electronics*, 1: 335–338.

MANN, H. M., 1967. Germanium gamma ray spectrometers. *Inst. Elec. Electron. Engrs., Trans. Nucl. Sci.*, 14: 10–26.

MARISCOTTI, M. A., 1967. A method for automatic identification of peaks in the presence of background and its application to spectrum analysis. *Nucl. Instr. Methods*, 50: 309–320.

MATHEVON, G., ROTHE, J. P., SIFFERT, P. et COCHE, A., 1967. Détermination de la radioactivité des roches par spectrométrie gamma à l'aide d'un détecteur au germanium compensé au lithium. *Compt. Rend.*, 264B: 1753–1756.

MAY, H. and MARINELLI, L. D., 1964. Cosmic ray contributions to the background of low level scintillation spectrometers. In: J. A. S. ADAMS and W. M. LOWDER (Editors), *The Natural Radiation Environment*. University of Chicago Press, Chicago, Ill., pp.463–483.

MAYER, J. W., 1966. Semiconductor detectors for nuclear spectrometry. *Nucl. Instr. Methods,* 43: 55–64.

MAZE, R., 1946. Compteurs à paroi de verre et à cathode externe. *J. Phys. Radium,* 7: 164–166.

MCNAIR, A., GLOVER, R. N. and WILSON, H. W., 1956. The decay of potassium-40. *Phil. Mag.*, 1: 199–211.

MEIER-LEIBNITZ, H., 1946. Absolute Zählrohrmessungen an gamma-Strahlung. *Z. Naturforsch.*, 1: 243–252.

MILLER, D. G., 1955. Proportional gamma-ray spectrometer. *Nucleonics*, 13, 2: 58–60.

MILLER, W. F. and SNOW, W. J., 1960a. Response spectra for gamma-rays in CsI. *Rev. Sci. Instr.*, 31: 861–862.

MILLER, W. F. and SNOW, W. J., 1960b. Energy loss spectra for gamma-rays in NaI. *Rev. Sci. Instr.*, 31: 39–45.

MISHRA, V. C., BRIAR, S. S. and GUSTAFSON, P. S., 1965. Calculation of the gamma-ray detection efficiency of annular and well type sodium iodide crystals. *Intern. J. Appl. Radiation Isotopes*, 16: 697–704.

MOLJK, A., DREVER, R. W. and CURRAN, S. C., 1955. Neutron activation applied to potassium mineral dating. *Nucleonics*, 13: 44–46.

MOORE, W. S. and SACKETT, W. M., 1964. Uranium and thorium series in equilibrium in sea water. *J. Geophys. Res.*, 69: 5401–5405.

MORGAN, J. W. and LOVERING, J. F., 1963. The determination of uranium and thorium in rocks by neutron activation analysis. *Anal. Chem. Acta*, 28: 405–417.

MOTT, W. E. and SUTTON, R. B., 1958. Scintillation and Cerenkov counters. In: *Handbuch der Physik*. Springer, Berlin, 45: 81–173.

MOXHAM, R. M., 1963. Natural radioactivity in Washington County, Maryland. *Geophysics*, 28: 262–272.

MÜLLER, D., 1964. Die Eigenschaften eines allseitig empfindlichen Proporzionalzählrohrs für die Röntgenfluoreszenanalyse. *Kernenergie*, 7: 249–254.

MURRAY, R. B., 1962. Scintillation counters. In: A. H. SNELL (Editor), *Nuclear Instruments and Their Use*. Wiley, New York, N.Y., pp.82–165.

NARBUTT, K. I., PEREL'MAN, S. M., PRAGER, I. A. and KHARLAKOV, V. A., 1963. Proportional counters for detecting X-rays and their use for X-ray spectroscopic analysis. *Izv. Akad. Nauk S.S.S.R., Ser. Phys.*, 27: 430–437.

NEILER, G. H. and BELL, P. R., 1965. The scintillation method. In: K. SIEGBAHN (Editor), *Alpha, Beta and Gamma Ray Spectroscopy, 1*. North-Holland, Amsterdam, pp.245–302.

NELMS, A. T., 1953. Graphs of the Compton energy–angle relationship and the Klein–Nishina formula from 10 kev to 100 Mev. *Natl. Bur. Std., (U.S.) Circ.*, 542.

NIELSEN, J. M. and PERKINS, R. W., 1967. Anticoincidence shielded multidimensional gamma-ray spectrometers for low level counting. In: *Radioactive Dating and Methods of Low Level Counting*. International Atomic Energy Agency, Vienna, pp. 687–702.

NIER, A. O., 1950. A redetermination of the relative abundances of the isotopes of carbon, nitrogen, oxygen, argon and potassium. *Phys. Rev.*, 77: 789–793.

NORLING, F., 1940. On the efficiency of gamma-ray counters. *Phys. Rev.*, 58: 277.

NURMIA, A., 1962. Experimental techniques in low level alpha spectrometry. In: *Proceedings, 1st Conference on Nuclear Electronics, Belgrade.* International Atomic Energy Agency, Vienna, pp.297–302.

OGAWA, I., DOKE, T. and TSUKUDA, M., 1961. The double grid ionization chamber. *Nucl. Instr. Methods*, 13: 169–176.

O'KEEFE, J. A., 1963. *Tektites.* University of Chicago Press, Chicago, Ill., 400 pp.

OVERSBY, V. M. and GAST, P. W., 1968. Lead isotope composition and uranium decay series disequilibrium in volcanic rocks. *Earth Planetary Sci. Letters*, 5: 199–206.

PELL, E. M., 1960. Ion drift in an *n–p* junction. *J. Appl. Phys.*, 31: 291–302.

PERKINS, R. W., THOMAS, C. W. and NIELSEN, J. M., 1965. Measurements of air-borne radionuclides and determination of their physical characteristics. In: *Proceeding of a Conference on Radioactive Fallout from Nuclear Weapons Tests, Washington.* U.S. At. Energy Comm., Washington.

PETTERSON, H., 1937. Der Verhältnis Thorium zu Uran in den Gesteinen und im Meer. *Mitt. Inst. Radiumforsch.*, 15, 400A.

PICKELL, J. J. and HEACOCK, J. G., 1960. Density logging. *Geophysics*, 25: 891–904.

PITKIN, J. A., NEUSCHEL, S. K. and BATES, R. G., 1964. Aeroradioactivity surveys and geological mapping. In: J. A. S. ADAMS and W. M. LOWDER (Editors), *The Natural Radiation Environment.* University of Chicago Press, Chicago, Ill., pp.723–736.

PRICE, B. T., MORTON, C. C. and SPEINNEY, K. C., 1957. *Radiation Shielding.* Pergamon, New York, N.Y., 350 pp.

PRICE, P. B. and WALKER, R. M., 1963a. Fossil tracks of charged particles in mica and the age of minerals. *J. Geophys. Res.*, 68: 4847–4862.

PRICE, P. B. and WALKER, R. M., 1963b. A simple method of measuring low uranium concentrations in natural crystals. *Appl. Phys. Letters*, 2: 23–25.

PRICE, W. J., 1964. *Nuclear Radiation Detectors.* McGraw-Hill, New York, N.Y., 430 pp.

PROSPERI, D. e SCIUTI, S., 1958. Su alcuni metodi radiometrici per la determinazione di Uranio et Torio nei minerali. *Ric. Sci.*, 28: 345–361.

PRUSSIN, S. G., HARRIS, J. A. and HOLLANDER, J. M., 1965. Application of lithium drifted germanium gamma-ray detectors to neutron activation analysis. *Anal. Chem.*, 37: 1127–1132.

PURVIS, A. E. and FOOTE, R. S., 1964. Atmospheric attenuation of gamma radiation. In: J. A. S. ADAMS and W. M. LOWDER (Editors), *The Natural Radiation Environment.* University of Chicago Press, Chicago, Ill., pp.747–756.

RAGLAND, P., 1964. Autoradiographic investigations of naturally occurring minerals. In: J. A. S. ADAMS and W. M. LOWDER (Editors), *The Natural Radiation Environment.* University of Chicago Press, Chicago, Ill., pp.129–151.

RAJEWSKI, B., 1943. Das Geiger–Müller Zählrohr im Dienste des Bergbaues. *Z. Phys.*, 120: 627–638.

RAMTHUN, H., 1964. Kalorimetrische Neubestimmung der Halbwertszeit von RaD (Pb-210). *Z. Naturforsch.*, 19a: 1064–1069.

RANKAMA, K., 1963. *Progress in Isotope Geology.* Interscience, New York, N.Y., 705 pp.

RAPOLLA, A. and VITTOZZI, P., 1968. Radioactivity of the vesuvian lavas and their dating by the disequilibrium ^{226}Ra–^{238}U. *Bull. Volcanol.*, 32, in press.

REED, G. W. and TURKEVICH, A., 1955. Uranium content of two iron meteorites. *Nature*, 176: 794–795.

REED, G. W., HAMAGUCHI, H. and TURKEVICH, A., 1958. The uranium content of iron meteorites. *Geochim. Cosmochim. Acta*, 13: 248–255.

REHMANN, W. und SCHARMANN, A., 1966. Zum Scintillazionsprozess in CsI(Tl). *Z. Naturforsch.*, 21A: 1684–1689.

REUTERSWARD, C., 1956. On the isotopic constitution of potassium. A study of mass discrimination in a hot anode ion source. *Arkiv Fysik*, 11: 1–54.

RICHARDSON, K. A., 1963. *Radioactivity, Sites of Alpha Emitters and Radioactive Disequilibrium in the Conway Granite of New Hampshire*. Thesis, Rice University, Houston, Texas.

RICHARDSON, K. A., 1964. Thorium, uranium and potassium in the Conway Granite, New Hampshire, U.S.A. In: J. A. S. ADAMS and W. M. LOWDER (Editors), *The Natural Radiation Environment*. University of Chicago Press, Chicago, Ill., pp.39–50.

RICHTER, H. G. and BALLARD, L. F., 1961. Construction and operating characteristics of flexible Geiger counters. *Nucl. Electronics*, 1: 339–344.

RICHTER, H. G. and GILLESPIE JR., A. S., 1961. Flexible Geiger counters. *U.S. At. Energy Comm., Res. Triangle Inst., Rept.*, ORO-482.

RITTMANN, A., 1933. Die geologische bedingte Evolution und Differentiation des Somma–Vesuv. magmas. *Z. Vulkan.*, 15: 8–45.

RITTMANN, A., 1948. Origine e differenziazione del magma ischitano. *Schweiz. Mineral. Petrog. Mitt.*, 28: 643–698.

ROGERS, J. J. W. and ADAMS, J. A. S., 1957. Autoradiography of volcanic rocks of Mount Lassen. *Science*, 125: 1150.

ROGERS, J. J. W. and ADAMS, J. A. S., 1963. Lognormality of thorium concentrations in the Conway Granite. *Geochim. Cosmochim. Acta*, 27: 775–783.

ROGERS, J. J. W. and ADAMS, J. A. S., 1969a. Thorium. In: *Handbuch der Geochemie*, Springer, Berlin, pp.90/1–90/0.

ROGERS, J. J. W. and ADAMS, J. A. S., 1969b. Uranium. In: *Handbuch der Geochemie*. Springer, Berlin, pp.92/1–92/0.

ROGERS, J. J. W., ADAMS, J. A. S. and GATLIN, B., 1965. Distribution of uranium, thorium and potassium concentrations in three cores from the Conway Granite, New Hampshire, U.S.A. *Am. J. Sci.*, 263: 817–822.

ROSE, M. E. and KORFF, S. A., 1941. An investigation of the properties of proportional counters. *Phys. Rev.*, 59: 850–859.

ROSHOLT, J. N., 1967. Open system model for uranium series dating of Pleistocene samples. In: *Radioactive Dating and Methods of Low Level Counting*. International Atomic Energy Agency, Vienna, pp.299–311.

ROSHOLT, J. N., SHIELD, W. P. and GARNER, E. L., 1963. Isotopic fractionation of uranium in sandstone. *Science*, 139: 224–225.

ROSHOLT, J. N., HARSHMAN, E. N., SHIELDS, W. R. and GARNER, E. L., 1964. Isotopic fractionation of uranium related to roll features in sandstone, Shirley Basin, Wyoming. *Econ. Geol.*, 59: 570–585.

ROSSI, B. and STAUB, J., 1949. *Ionization Chambers and Counters*. McGraw-Hill, New York, N.Y., 243 pp.

ROUBAULT, M., DE LA ROCHE, H. et GOVINDARAJU, K., 1966. Rapport sur quatre roches étalons géochimiques: granites GR, GA, GH and basalt BR. *Sci. Terre*, 11: 105–121.

ROWE, M. W., VAN DILLA, M. A. and ANDERSON, E. C., 1963. On the radioactivity of stone meteorites. *Geochim. Cosmochim. Acta*, 27: 983–1000.

ROY, R. F., BLACKWELL, D. D. and BIRCH, F., 1968. Heat generation of plutonic rocks and continental heat flow provinces. *Earth Planetary Sci. Letters*, 5: 1–12.

RYBACH, L. and ADAMS, J. A. S., 1966. Automatic analysis of the elements U, Th and K in solid rock samples by nondestructive gamma spectrometry. *Proc. Anal. Chem. Conf. Budapest*, pp.323–330.

RYBACH, L., VON RAUMER, J. and ADAMS, J. A. S., 1966. A gamma spectrometric study of Mont Blanc granite samples. *Pure Appl. Geophys.*, 63: 153–160.

SACKETT, W. M., PORTRATZ, H. and GOLDBERG, E. D., 1958. Thorium content of ocean water. *Science*, 128: 204–205.

SAKANOUE, M., KONISHI, K. and KAMURA, K., 1967. Stepwise determination of thorium, protactinium and uranium isotopes and their applications in geochronological studies. In: *Radioactive Dating and Methods of Low Level Counting*. International Atomic Energy Agency, Vienna, pp.313–330.

SALVINI, G., 1948. Some Geiger–Müller and proportional counters of spherical shapes. *Rev. Sci. Instr.*, 19: 494–496.

SAVELLI, C., 1967a. The problem of rock assimilation by Somma–Vesuvius magma: I. Composition of Somma and Vesuvius lavas. *Contrib. Mineral. Petrog.*, 16, 328–363.

SAVELLI, C., 1967b. The problem of rock assimilation by Somma–Vesuvius magma: II. Composition of sedimentary rocks and carbonate ejecta from the Vesuvius area. *Contrib. Mineral. Petrog.*, 18: 43–64.

SCHAEFFER, O. A., 1962. Radiochemistry of meteorites. *Ann. Rev. Phys. Chem.*, 13: 161–168.

SCHAEFFER, O. A., 1966. Tektites. In: O. A. SCHAEFFER and J. ZÄHRINGER (Editors), *Potassium–Argon Dating*. Springer, New York, N.Y., pp.162–173.

SCHLECHT, W. G. and STEVENS, R. E., 1951. Results of chemical analyses of samples of granite and diabase. *U.S. Geol. Surv. Bull.*, 980: 7–24.

SCHONFELD, E., KIBBEY, A. H. and DAVIS JR., W., 1966. Determination of nuclide concentrations in solutions containing low level radioactivity by least square resolution of the gamma-ray spectra. *Nucl. Instr. Methods*, 45: 287–292.

SCHNEIDER, H. and SCHWERDTEL, E., 1960. Portable scintillometer with a differential discriminator. *Atomkernenergie*, 5: 278–281.

SCHREINER, G. D. L. and VERBEEK, A. A., 1965. Variations in $^{39}K/^{41}K$ ratio and movement of potassium in a granite-shale contact region. *Proc. Roy. Soc. London, Ser. A*, 285: 423–429.

SCOTT, J. H., DODD, P. H., DROULLARD, R. F. and MUDRA, P. J., 1961. Quantitative interpretation of gamma-ray logs. *Geophysics*, 26: 182–191.

SEGRE', E., 1965. *Nuclei and Particles. An Introduction to Nuclear and Subnuclear Physics*. Benjamin, New York, N.Y., 741 pp.

SEYFANG, A. P., 1955. An improvement in the determination of uranium-235 by radioactivation. *Analyst*, 80: 74–80.

SEYFANG, A. P. and SMALES, A. A., 1953. The determination of uranium-235 in mixtures of naturally occurring uranium isotopes by radioactivation. *Analyst*, 78: 394–405.

SHIRLEY, D. A., 1965. Applications of germanium gamma-ray detectors. *Nucleonics*, 23: 62–69.

SHORT, M. A., 1960. Detection and correction of nonlinearity in X-ray proportional counters. *Rev. Sci. Instr.*, 31: 618–620.

SIEGBAHN, K. (Editor), 1965. *Alpha, Beta and Gamma Ray Spectroscopy, 1*. North-Holland, Amsterdam, 1800 pp.

SIEVERT, I. und SCHMIDT, H., 1966. Zur Spektrometrie weicher Röntgenstrahlung in Proportionalzählern, I, II. *Nucl. Instr. Methods*, 42: 45–55.

SIKOSEK, B. and GOJKOVIC, S. E., 1966. An experimental determination of geochemical facies of sedimentary rocks in the Dinarides based on Th/U ratio. *Freiberger Forschungsh., C*, 210: 197–211.

SILK, E. C. and BARNES, R. S., 1959. Examination of fission fragments tracks with an electron microscope. *Phil. Mag.*, 4: 970–972.

SINE, N. M., TAYLOR, W. O., WEBBER, G. R. and LEWIS, G. L., 1969. Third report of analytical data for CAAS sulphide ore and syenite rock standards. *Geochim. Cosmochim. Acta*, 33: 121–132.

SLIV, L. A. and BOND, I. M., 1965. Tables of internal conversion coefficients. In: K. SIEGBAHN (Editor), *Alpha, Beta and Gamma Ray Spectroscopy, 1.* North-Holland, Amsterdam, pp.1639–1672.

SMALES, A. A., 1952. The determination of small quantities of uranium in rocks and minerals by radioactivation. *Analyst*, 77: 778–789.

SNYDER, B. J., 1967. Tabulated efficiencies of some well-type gamma-ray scintillation detectors for various geometries. *Nucl. Instr. Methods*, 53: 313–319.

SNYDER, B. J. and GYOREY, G. L., 1965. Calculating gamma-ray efficiencies in scintillation detectors. *Nucleonics*, 23: 80–86.

SNYDER, B. J. and KNOLL, G. F., 1966. Calculated gamma-ray photofractions for well type scintillation detectors. *Nucl. Instr. Methods*, 40: 261–266.

SOMAYAJULU, B. L. K., TATSUMOTO, M., ROSHOLT, J. N. and KNIGHT, R. J., 1966. Disequilibrium of the ^{238}U series in basalts. *Earth Planetary Sci. Letters*, 1: 387–391.

SONNTAG, C., 1967. Extremely low level scintillation spectrometer. In: *Radioactive Dating and Methods of Low Level Counting.* International Atomic Energy Agency, Vienna, pp.675–685.

SOROIOU, M., CEREI, M., ONCESCU, M., DANIS, A. and MANTESCU, C., 1965. Radiogenic argon determination by neutron activation analysis. *Geochim. Cosmochim. Acta*, 29: 551–561.

SPIERS, F. W., 1950. Radioactivity of potassium. *Nature*, 165: 356.

SPURNY, Z., 1965. Thermoluminescent dosimetry. *At. Energy Rev.*, 3: 61–115.

STEVENS, R. E., 1960. Second report on cooperative investigation of the composition of two silicate rocks. *U.S. Geol. Surv. Bull.*, 1113: 1–126.

STEVER, H. G., 1942. The discharge mechanism of fast GM counters from the dead time experiments. *Phys. Rev.*, 61: 38–52.

STOENNER, R. W. and ZAHRINGER, J., 1958. Potassium–argon age of iron meteorites. *Geochim. Cosmochim. Acta*, 15: 40–50.

STRAUSS, M. C., SHERMANN, L. S., BRENNER, S., RODNICK, S. J., LARSEN, R. N. and MANN, H. M., 1967. High resolution Ge(Li) spectrometer for high input rates. *Rev. Sci. Instr.*, 38: 725–730.

SUTTLE JR., A. D. and LIBBY, W. F., 1955. Absolute essay of beta radioactivity in thick solids. Application to naturally radioactive potassium. *Anal. Chem.*, 27: 921–927.

SUZUKI, T. and YUHARA, J., 1961. Counting efficiency of gamma-rays in counting tubes. *J. Phys. Soc. Japan*, 16: 152–156.

SYROMYATNIKOV, N. G., IBRAEV, R. A. and MUKASHEV, S. A., 1967. The interpretation of uraniummetrical anomalies in arid regions using the isotopic ratio $^{234}U/^{238}U$. *Geochim. Intern.*, 7: 834–841.

TADDEUCCI, A., BROECKER, W. S. and THURBER, D. L., 1967. ^{230}Th dating of igneous rocks. *Earth Planetary Sci. Letters*, 3: 338–342.

TAVENDALE, A. J., 1965. Large germanium–lithium drift *p–i–n* diodes for gamma-ray spectroscopy. *Inst. Elec. Electron. Engrs., Trans. Nucl. Sci.*, 12: 255–264.

TAVENDALE, A. J., 1966. Characteristics of some large coaxial Lithium drift semiconductor gamma ray spectrometers. *Inst. Elec. Electron. Engrs., Trans. Nucl. Sci.*, 13: 315–327.

TAYLOR, J. and PARRISH, W., 1955. Absorption and counting efficiency data for X-ray detectors. *Rev. Sci. Instr.*, 26: 367–373.

TAYLOR, J. M., 1963. *Semiconductor Particle Detectors.* Butterworths, Washington, D.C., 180 pp.

THURBER, D. L., 1962. Anomalous $^{234}U/^{238}U$ in nature. *J. Geophys. Res.*, 67: 4518–4520.

TITTMANN, J. and WAHL, J. S., 1965. The physical foundations of formation density logging (gamma–gamma). *Geophysics*, 30: 284–294.

TIWARI, P. N. and KONDAIAH, E., 1966. Total and photopeak efficiencies for a well type NaI(Tl) crystal for cylindrical sources. *Nucl. Instr. Methods*, 42: 118–124.

TOBAILEM, J., 1955. Mesure de la periode du ThB (^{212}Pb). *J. Phys. Radium*, 16: 115–118.

TOBAILEM, J., DAVID, B. et NORDEMANN, D., 1967. Radioactivité induite par le rayonnement cosmique dans le météorite Saint Severin. En: *Radioactive Dating and Methods of Low Level Counting.* International Atomic Energy Agency, Vienna, pp.207–213.

TOGLIATTI, V., 1965a. Curve di saturazione per fissioni fossili di ^{238}U in muscovite. *Boll. Geofis. Teor. Appl.*, 7: 251–260.

TOGLIATTI, V., 1965b. Distribuzione dei ranges e distribuzione angolari delle tracce di fissione fossile di ^{238}U e di fissione indotta di ^{235}U in mica. *Boll. Geofis. Teor. Appl.*, 7: 326–335.

TOLCHENOV, Y. M., 1960. Effect of the gamma-ray background on gas amplification factor in a proportional counter. *Pribory i Tekn. Eksperim.*, 3: 38–41.

TROST, A., 1958. Röntgeninterferenzmessungen mit Proportional–Zählrohr und Einkanal Diskriminator. *Z. Angew. Phys.*, 10: 404–410.

UREY, H. C., 1955. The cosmic abundances of potassium, uranium and thorium and the heat balances of the earth, the moon and Mars. *Proc. Natl. Acad. Sci.*, 41: 127–144.

VAN DILLA, M. A., ARNOLD, J. R. and ANDERSON, E. C., 1960. Spectrometric measurements of natural and cosmic ray induced radioactivity in meteorites. *Geochim. Cosmochim. Acta*, 20: 115–121.

VAN LIESHOUT, R., WAPSTRA, A. H., RICCI, R. A. and GIRGIS, R. K., 1965. Scintillation spectra analysis. In: K. SIEGBAHN (Editor), *Alpha, Beta and Gamma-Ray Spectroscopy, 1.* North-Holland, Amsterdam, pp.501–537.

VERBEEK, A. A. and SCHREINER, G. D. L., 1967. Variations in ^{39}K: ^{41}K ratio and movement of potassium in a granite-amphibolite contact region. *Geochim. Cosmochim. Acta*, 31: 2125–2134.

VINOGRADOV, A. P., SURKOV, Y. A., CHERNOV, G. M., KIRNOZOV, F. F. and NAZARKINA, G. B., 1966. Measurements of gamma radiation from the surface of the moon by the cosmic station "Luna 10." *Geochim. Intern.*, 3: 707–715.

VITTOZZI, P., 1964. Spettrometria alfa di rocce radioattive. *Ann. Osserv. Vesuviano*, S.VI, 6: 85–112.

VITTOZZI, P. e GASPARINI, P., 1963. Determinazione della concentrazione di isotopi radioattivi nelle rocce mediante la spettrometria gamma. *Atti Conv. Assoc. Geofis., Ital., 12th, Roma*, pp.219–236.

VITTOZZI, P. and GASPARINI, P., 1964. Gamma ray spectra of some lavas from Vesuvius. *Bull. Volcanol.*, 27: 301–314.

WANKE, H. und KONIG, H., 1959. Eine neue Methode zu Kalium–Argon Alterbestimmung und ihre Anwendung auf Steinmeteorite. *Z. Naturforsch.*, 14a: 860–866.

WAPSTRA, A. H., 1965. The coincidence method. In: K. SIEGBAHN (Editor), *Alpha, Beta, and Gamma Ray Spectroscopy, 1.* North-Holland, Amsterdam, pp.539–555.

WAPSTRA, A. H., NIJGH, G. J. and VAN LIESHOUT, R., 1959. *Nuclear Spectroscopy Tables.* North-Holland, Amsterdam, 135 pp.

WASSERBURG, G. J. and HAYDEN, R. J., 1954. The branching ratio of K^{40}. *Phys. Rev.*, 93: 645.

WASSERBURG, G. J., MACDONALD, G. J. F., HOYLE, F. and FOWLER, W. A., 1964. The relative contributions of uranium, thorium and potassium to heat production in the earth. *Science*, 142: 465–467.

WASSON, J. T., ALDER, B. and OESCHGER, H., 1967. Aluminum-26 in Pacific sediments: implications. *Science*, 155: 446–448.

WATT, D. E. and RAMSDEN, D., 1964. *High Sensitivity Counting Techniques.* Pergamon, London, 348 pp.

WEBB, P. P., GREEN, R. M., FOWLER, L. L. and MALM, H. L., 1966. Encapsulated germanium gamma ray spectrometers. *Inst. Elec. Electron. Engrs., Trans. Nucl. Sci.,* 13: 351–358.

WEBBER, G. R., 1965. Second report of analytical data for CAAS syenite and sulphide standards. *Geochim. Cosmochim. Acta*, 29: 229–248.

WELLER, R. I., 1964. Low level radioactive contamination. In: J. A. S. ADAMS and W. H. LOWDER (Editors), *The Natural Radiation Environment.* University of Chicago Press, Chicago, Ill., pp.567–576.

WEST, D., 1953. Energy measurements with proportional counters. In: O. R. FRISCH (Editor), *Progress in Nuclear Physics, 3.* Pergamon, London, pp.18–62.

WEST, D. and ROTHWELL, P., 1950. Proportional counters in a magnetic field: an investigation of the isomerism of ^{80}Br and a measurement of the fluorescence yield of krypton. *Phil. Mag.*, 43: 873–899.

WEST, D., DAWSON, J. K. and MANDLEBERG, C. J., 1952. The investigation of soft radiations from ^{239}Pu and ^{233}U with a proportional counter. *Phil. Mag.*, 45: 875–888.

WETHERILL, G. W., 1957. Radioactivity of potassium and geological time. *Science*, 126: 545–546.

WETHERILL, G. W. and TILTON, G. R., 1967. Geochronology. In: P. H. ABELSON (Editor), *Researches in Geochemistry, 2.* Wiley, New York, N.Y., pp.1–28.

WETHERILL, G. W., TILTON, G. R., DAVIS, G. R. and ALDRICH, L. T., 1956a. New determinations of the age of the Bob Ingersoll pegmatite, Keystone, South Dakota. *Geochim. Cosmochim. Acta*, 9: 292–297.

WETHERILL, G. W., WASSERBURG, G. J., ALDRICH, L. T., TILTON, G. R. and HAYDEN, R. J., 1956b. Decay constants of ^{40}K as determined by the radiogenic argon content of potassium minerals. *Phys. Rev.*, 103: 987–989.

WHITE, F. A., COLLINS, T. L. and RURKE, F. M., 1956. Search for possible naturally occurring isotopes of low abundance. *Phys. Rev.*, 101: 1786–1791.

WILKINSON, D. H., 1950. *Ionization Chambers and Counters.* Cambridge University Press, Cambridge, Mass., 266 pp.

WILSON, R., BEGHIAN, L., COLLIE, C. H., HALBAN, H. and BISHOP, G. R., 1950. High pressure ionization chambers and their use. *Rev. Sci. Instr.*, 21: 609–705.

WING, J. and WAHLGREN, M. A., 1967. Detection sensitivities in nuclear activation with an isotopic fast neutron source. *Anal. Chem.*, 39: 85–89.

WOLLENBERG, H. A. and SMITH, A. R., 1964. Studies in terrestrial gamma radiation. In: J. A. S. ADAMS and W. M. LOWDER (Editors), *The Natural Radiation Environment*. University of Chicago Press, Chicago, Ill., pp.513–566.

WOLLENBERG, H. A. and SMITH, A. R., 1968. Radiogeologic studies in the central part of the Sierra Nevada Batholith, California. *J. Geophys. Res.*, 73: 1481–1495.

YAGODA, H., 1949. *Radioactive Measurements with Nuclear Emulsions*. Wiley, New York, N.Y., 450 pp.

YOKOYAMA, Y., TOBAILEM, J., GRJEBINE, T. et LABEYRIE, J., 1968. Détermination de la vitesse de sédimentation océanique par une méthode non destructive de spectrométrie gamma. *Geochim. Cosmochim. Acta*, 32: 347–352.

YOUNG, M. H. and SINGHAL, M. S., 1966. Gamma ray spectra unfolding by the stepwise statistical method. *Nucl. Instr. Methods*, 45: 287–292.

ZASTAWNY, A., 1966. Gas amplification in a proportional counter with carbon dioxide. *J. Sci. Instr.*, 43: 179–181.

ZASTAWNY, A., 1967. On a new formula for the gas amplification in a proportional counter. *J. Sci. Instr.*, 44: 395.

ZASTAWNY, A. and MIZERACZYK, J., 1966. Gas amplification in proportional counters with argon and nitrogen. *Nukleonika*, 11: 685–689.

ZELLER, E. J., 1954. Thermoluminescence of carbonate sediments. In: H. FAUL (Editor), *Nuclear Geology*. Wiley, New York, N.Y., pp.180–188.

ZELLER, E. J., LEVY, P. W. and MATTERN, P. L., 1967. Geological dating by electron spin resonance. In: *Radioactive Dating and Methods of Low Level Counting*. International Atomic Energy Agency, Vienna, pp.531–540.

ZERBY, C. D. and MORAN, H. S., 1961. Calculation of the pulse-height response of NaI(Tl) scintillation counters. *Nucl. Instr. Methods*, 14: 115–124.

ZIEMBA, F. P., PELT, G., RYANG, G., WANG, L. and ALEXANDER, R., 1962. Properties of an $n^+ i\, p^+$ semiconductor detector. *Inst. Radio Engrs., Trans. Nucl. Sci.*, 9: 155–159.

Index